Gravity's Time

C. S. Unnikrishnan

CRC Press
Taylor & Francis Group
Boca Raton London New York

CRC Press is an imprint of the
Taylor & Francis Group, an **informa** business

Gravity's Time

First published 2022 by Jenny Stanford Publishing Pte. Ltd.

Published 2023 by CRC Press
Taylor & Francis Group
6000 Broken Sound Parkway NW, Suite 300
Boca Raton, FL 33487-2742

ISBN 13: 978-981-4877-72-5 (hbk)

Visit the Taylor & Francis Web site at
http://www.taylorandfrancis.com

and the CRC Press Web site at
http://www.crcpress.com

British Library Cataloguing-in-Publication Data
A catalogue record for this book is available from the British Library.

Remembering My Grandfather,

Who Taught Me How to Read Time

Contents

List of Figures

List of Tables

Preface

This book is a mission to give gravity its due. There is supposedly a complete theory of gravity of which we celebrated the centenary year in 2016. Einstein's General Theory of Relativity (GTR) could very well have been the last word on gravity at our familiar scales, but for the fact that the theory was ahead of the times, taking premature birth. Like its predecessor – the Special Theory of Relativity (STR) – it is built on the basis of an *empty space*, with negligible matter content, as its fundamental premise. The characteristic feature of the STR, stated as its fundamental postulate, is the invariance of the relative velocity of light in any inertial frame. Its most notable physical effect is time dilation – the slowing down of the rate of clocks in relative motion. In addition, the GTR has the gravitational time dilation; clocks tick slower in a gravitational field. The physical effects due to gravity are calculated in the GTR with matter added to the pre-existing empty space. If one thinks hard enough, one can easily see that time never could exist in pure empty space. While one may imagine empty space in our fertile mind-space, clocks and time do not exist without matter. What we call time is factually the change in the physical state of some matter, in all cases imaginable. Time cannot be separated from matter, and matter without any evolution of its physical state cannot mark time. This change could be a simple motion of matter relative to another piece of matter, but some change in the physical state of being is essential to mark time. Time is material embedding in space. So, *empty space is timeless.*

Yet, one may argue that this is not a serious gap in our theories because we can assume that there is just enough matter to mark our time and that should not make any difference to the theory itself. Nothing could be farther from the fact and truth.

Both the Special Theory of Relativity and the General Theory of Relativity were completed and indeed accepted as plausibly the correct theories well before we knew anything significant about the matter around us in the Universe. Therefore, *the factual gravity of the vast amounts of matter in the matter-filled Universe is absent in these theories.* These theories took birth, and grew, when we remained ignorant about the matter content of the vast Universe. The result is that we ended up with a running cart with no horse to spot.

Special Relativity's clocks and time are modifiable but that has nothing whatsoever to do with gravity, or with any physical interaction. Special Relativity's clocks in *relative motion* change their rates without any physical reason. In contrast, General Relativity predicted the relative time dilation for clocks in different gravitational potentials. Yet, the theory missed the largest of gravitational potential we experience – the cosmic gravitational potential itself! As constructed, the GTR is operative in this Universe, in the presence of the cosmic matter, and yet the theory ignores cosmic gravity in its structure. A moving clock in the matter-filled space will surely experience a different potential compared to a stationary clock, unlike in empty space. The effects are large, if one bothers to calculate. Stated differently, if the theories of motion and gravity were formulated after the vast Universe and its contents became visible to us through the mighty telescopes that became operational after 1920, the history of physics would have been very different, and our physics would have been closer to truth. So, either we have made a big error and all motional effects are indeed gravitational, involving all cosmic matter, or we have to assert that the enormous amount of cosmic matter we observe with our large telescopes has no gravity. The choice is obvious.

The correct choice brings with it a revolutionary revision of our understanding of time. If cosmic gravity is the determining factor, then the cosmic frame is also a master preferred frame. It makes a difference whether one is moving relative to the cosmic matter or staying at rest. Therefore, *the relative velocities between two clocks could not be the criterion for deciding time dilations – it has to be the velocity relative to the cosmic frame.* We are, of course, back to the absolute space and time that existed in our physics and philosophy. The radical change in the physical paradigm is obvious. It is easy to verify that all empirical results are not only consistent with this choice, but also are demanded by many a factual evidence. There is more than enough evidence in clock data, including the GPS, to show that what is relevant is not the velocity relative to the observer's rest frame, but the velocity relative to the cosmic frame. The navigational aids in the smart phones we carry are 'pocket proofs' that the prevailing theories of time need a drastic modification.

We will see that in fact there is an absolute clock synchronized all over the Universe. The temperature of the cosmic microwave background radiation is the same everywhere to a part in million or better, once the known small fluctuations are averaged out. This temperature is monotonically decreasing, providing us with the absolute 'cosmic clock'. It selects out one unique sequence of time slices in the entire Universe; then the general coordinate invariance that is one of the pillars of General Relativity is broken in the real Universe.

With this realization, Einstein's time goes out of the window of the vast edifice of physics. In fact, the situation in the physics of clocks, time, and relativity is more fluid than anybody could have suspected. The subtlety of the important issue of synchronizing clocks at different locations, foreseen and analyzed by Poincaré, became a fatal misinterpretation in the STR. This single crucial error, uncorrected despite an early warning critique by the philosopher Henri Bergson, remained with this theory and its history, forcing experimenters and GPS scientists to defend repeated anomalies with poor logic. Identifying cosmic gravity as the sole reason for the relativistic modification of time puts an end to a century of embarrassment of defending the inadequate physical theory. There is a special class of frames that are at rest relative to the cosmic matter, and all other frames see a matter current proportional to their velocity and hence a spatial anisotropy. Lorentz invariance, being violated by the real Universe, cannot be the basis of correct physics in spite of its accidental success, because most considerations depend only on the second-order part of the transformations (time dilation and length contraction) that are also part of the relativity theories with a preferred frame (like the ether, if one remembers the Lorentz–Poincaré theory).

With the absolute time and space back in the structure of physics, and with the current theory of relativity shown as inadequate, we are discussing time in a new paradigm. The gravitational paradigm of relativity correctly acknowledges the reality of cosmic gravity and incorporates it in the new theory of Cosmic Relativity. We do not discuss in this book all the aspects of the modified theory of relativity that I have developed in detail, except the connections relevant and essential for the discussion of time. The complete theory of Cosmic Relativity, its experimental support, and the induced modification of General Relativity are discussed in another monograph and in numerous papers. Cosmic Relativity is fully consistent with all known experimental evidence and it provides the most accurate description of the motional modification of time.

Time is gravity's tracker and faithful shadow. Once the frequency of an oscillator is specified, its phase is identical to time. But 'frequency' is akin to the energy in the quantum regime, and energy and mass are the sources of gravity. Energy being the 'charge' of gravity, time is affected by gravity in a universal way. The consistency circle is completed in Cosmic Relativity.

This book is not a discussion on the general nature of time within physics, or on the philosophical questions concerning time. Yet, we will touch on some relevant aspects briefly, when needed for clarity. These clarifications will have to necessarily be limited to the nature of the physical time.

It is high time that cosmic gravity got its rightful due as the determining factor for dynamics, relativity, and as the master of time. After centuries of search, we have mature and ample empirical evidence, all very transparent and direct, to guide and convince us about the true relation between cosmic gravity and TIME, and through that, about the intimate relation between the vast cosmos and us. We can now understand time in its true physical sense and completeness.

IT IS GRAVITY'S TIME.

Mumbai and Paris
December 2021

Acknowledgments

A monograph cannot just take its birth, with only its author in a role of creation. It needs an environment of both intellectual and emotional richness. In that I did not lack, because of the many colleagues and friends at the Tata Institute of Fundamental Research, Mumbai, where I worked ever since my graduate days, and in other laboratories where I was a frequent visitor. Looking back, it is wonderful to see that the whole of India, and beyond, was my land of activity in physics, and I must have gained plenty of wisdom from the numerous conversations on physics, art, and philosophy, without always realizing their influences.

As an unforgettable representative of the spirit of free mind and inquiry, I want to remember 'Rad', who was engaged in a conversation with me on some of the topics discussed in this book. I miss Prof. V. Radhakrishnan and the help I would have received from him in advancing true reasoned science. This book does not have a foreword because he is no more with us to write it.

My parents and my sister Karthika, keen observers of the progress of my work in fundamental physics for years, and my brother Hari, with his unconditional trust, have been providing the kind of support only family can.

Martine Armand was the constant companion all throughout the preparation, writing, arranging, and completing of this work. She did extensive scrutiny of the structure and presentation, and brought the final result in a vastly improved form.

Arvind Kanswal and the editorial team diligently ensured the progress of the book.

Finally, I thank my patient publishers, Stanford Chong and Jenny Rompas, who were very generous with time!

Chapter 1

Time

We explore the concept of Time with a desire to grasp it. . .

Then we realize that we can't take even our baby steps on this road because we do not have the ingredients to make 'time'. Our abstract time should be the distilled essence of our experienced time, as it is the case of all other physical quantities. And what do we see when we search our experience for the definition and meaning of time? Words that characterize time are 'Evolution', 'Change', 'Order', 'Memory', etc. But, all these notions dissolve away outside our consciousness, in the empty physical space. Pure emptiness cannot change. It cannot evolve, and it has no order or memory. *Empty space does not have time.* All our analyses of time finally focus on the evolution of the physical states of matter. Dynamics of matter suggests the concept of time, in its elementary nature. Motion, growth and decay, changes of all kinds, are what seed and reinforce our notion and experience of time. The seat of all these is matter and its energy distributed in space. The manifestation of time is only through the medium of matter.

We may at least pretend, with our vast experience, that we understand space and matter because we *can* imagine space without time or matter, and matter itself is our daily experience (the act

Gravity's Time
C. S. Unnikrishnan

ISBN 978-981-4877-72-5 (Hardcover), 978-1-003-25656-4 (eBook)
www.jennystanford.com

of 'imagining' requires the gray matter called the mind, but we may ignore that for the time being due to our necessity to start somewhere). Time as we know is inseparable from our notions of matter. *However hard we try, there is no viable notion of time in space devoid of all matter.* We should not be surprised or despaired – Newton had tried to reach to the self of time, by defining it as a *self-evident* element of nature, but even he could not go beyond a theological soliloquy. This attempt was just a repetition of what he tried for defining absolute space. However, realizing the need to define tangible concepts of space and time that were fundamental to the dynamics he was developing, he defined usable space and time as 'relative', from the material reference markers and the evolution of matter.

The inseparable relation between matter and time may suggest, but not establish, a relation between gravity and time, because there is no matter without gravity. All matter, whether it is in a form tangible to the senses or in the form of energy of various kinds, generates, and also responds to, gravity. However, the relation between gravity and time is subtle, and hidden in the very core of physics. Theoretical developments like the General Theory of Relativity brought out this connection to some extent, but the actual roots go very deep. Today, 'keeping time' means much more in life than ever. Without our direct awareness, the deep and subtle relation between matter, gravity, and time is guiding and affecting the course of our lives, as manifested in daily navigation, banking, travel, safety, etc.

Thus, we realize that to understand time in a way we can use it in our lives with definite mastery and control, we need to understand it physically. The physical understanding of time enriches its psychological and intuitive understanding, revealing the harmony of our total empirical experience. Even the many layers of the philosophical understanding of time come within our grasp with the correct physical understanding of time. After all, marking and measuring time have been one of the most discussed tasks assigned by civilizations and human society to physicists and natural philosophers.

1.1 Space

We need to discuss the concept of 'space' and understand it before we can explore the nature of time. Though time is very different from space, in experience, time is realized and felt in space. Time is measured in space, with material devices occupying space.

We feel that we understand space better than time. Perhaps that is right, because space is an extension of ourselves. In fact, it is extension itself, and therefore space extends with our imagination to infinities without facing conceptual problems. Our visual reach covers both near and far; as near as our bodies and as far as the stars. However, our familiarity of space is limited to three dimensions. All discussions on a larger number of spatial directions have remained as speculations, without even any indirect evidence for their plausibility. Thus, *space in real physics today has the same number of dimensions as matter itself*, and extended matter is the guide for our notions of space. In fact, if one thinks how we came to the conclusion that space is of three dimensions, the only reasonable answer is that we know only the dimensions of matter, and both the notion and the dimensions of space are the extrapolations of the properties of matter.

Newton attempted a description of ideal space, as an entity that was prior to matter, perhaps because he was convinced of its existence. Newton's approach for a definition of space and time is similar to Aristotle's approach in *Physics: Book IV*, the earliest of such an attempt explicitly in the context of the physics of motion (Aristotle, 1991). However, Newton made much progress compared to Aristotle. The real advance was in trying to distinguish between the 'relative' and the 'absolute', as we will see. Aristotle, as well as Galileo much later, acknowledged motion as sensible only relative to other things, whereas Newton recognized that all motion cannot be reduced to a single category.

In spite of Newton's general statement in his *Principia Mathematica* that the terms 'space', 'time' and 'motion' are well known to all, he put down formal definitions for the fundamental notions of space and time as the most essential a priori elements for the description

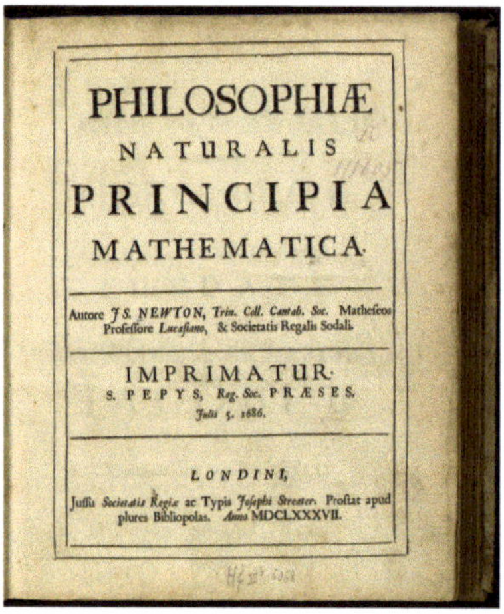

Figure 1.1 *Newton's Principia* (1686), the basis of the theory of dynamics.

of dynamics (Newton, 1846). In doing so, he clearly distinguished between the 'relative' and 'absolute' nature of motion, a division that has lasted to this day, dividing physical theories as well as philosophies sharply. He even provided an experimental proof for the reality of the 'absolute space', which remained unchallenged till two centuries later. Let us briefly examine the relevant aspects of Newtonian space and time.

Newton's concepts of categories of space and time are defined and described in detail in the first 'scholium' of Book 1 of his Principia (Fig. 1.1), after definitions, and before the 'axioms' on the laws of motion.

> SCHOLIUM
>
> Hitherto I have laid down the definitions of such words as are less known, and explained the sense in which I would have them to be understood in the following discourse. I do not define time, space, place and motion, as being well known to all. Only I must observe, that the vulgar conceive those quantities under no other notions but from the relation they bear to sensible objects. And thence arise certain prejudices, for the removing of which, it will be convenient to distinguish them into absolute and relative, true and apparent, mathematical and common.

His well-known assertions of time and space follow:

> I. Absolute, true, and mathematical time, of itself, and from its own nature flows equably without regard to anything external, and by another name is called duration: relative, apparent, and common time, is some sensible and external (whether accurate or unequable) measure of duration by the means of motion, which is commonly used instead of true time; such as an hour, a day, a month, a year.
>
> II. Absolute space, in its own nature, without regard to anything external, remains always similar and immovable. Relative space is some movable dimension or measure of the absolute spaces; which our senses determine by its position to bodies; and which is vulgarly taken for immovable space; such is the dimension of a subterraneous, an aereal, or celestial space, determined by its position in respect of the earth. Absolute and relative space, are the same in figure and magnitude; but they do not remain always numerically the same. For if the earth, for instance, moves, a space of our air, which relatively and in respect of the earth remains always the same, will at one time be one part of the absolute space into which the air passes; at another time it will be another part of the same, and so absolutely understood, it will be perpetually mutable. . .
>
> But we may distinguish rest and motion, absolute and relative, one from the other by their properties, causes and effects. It is a property of rest, that bodies really at rest do rest in respect to one another. And therefore as it is possible, that in the remote regions of the fixed stars, or perhaps far beyond them, there may be some body absolutely at rest; but impossible to know, from the position of bodies to one another in our regions whether any of these do keep the same position to that remote body; it follows that absolute rest cannot be determined from the position of bodies in our regions.

Newton then prescribes *how to distinguish empirically between absolute and relative motion*, motion relative to the absolute space and relative to other bodies. His 'water and pail' experiment has been discussed and debated by physicists and philosophers over the next three centuries, without consensus. We do not need to discuss these details. It will suffice to realize that Newton's distinction between absolute and relative space was the appearance and experience of the centrifugal force only when bodies rotate with respect to the 'absolute space'. Two centuries later, the philosopher-physicist Ernst Mach expressed his remarkable criticism of Newton's dependence on the insensible and metaphysical absolute space, as the basis for the physical explanation of the centrifugal and inertial forces (Mach, 1919). Mach suggested that some physical interaction with the stars, nebulae and the rest of the matter in the Universe must be the cause for the inertial forces. It is an empirical fact that such forces appear when the rotation is relative to the celestial frame of distant motionless stars. However, with very little matter seen and known then, the gravitational interaction (or any known interaction) could not be explicitly invoked as the reason. This was because the law of gravity was well measured and characterized, and the potentials and forces due to the matter visible and known then would have been too small to be taken seriously as the source of large inertial forces, despite the strong clue that such forces were 'inertial', proportional to the mass, just as gravity was. During the discussion of Newton's notions of space and the inertial forces, Mach wrote in his 'Science of Mechanics' something deep and significant about time as well:

> When we say a thing A changes with the time, we mean simply that the conditions that determine a thing A depend on the conditions that determine another thing. The vibrations of a pendulum take place in time when its excursion depends on the position of the earth. Since, however, in the observation of the pendulum, we are not under the necessity of taking into account its dependence on the position of the earth, but may compare it with any other thing (the conditions of which of course also depend on the position of the earth), the illusory notion easily arises that all the things with which we compare it are unessential. Time, accordingly, appears to be some particular and independent thing, on the progress of which the position of the pendulum depends, while the things that we resort to for comparison and choose at random appear to play a wholly collateral part. But we must not forget that all things in the world are connected with one another and depend on one another. . .

Note the particularly important point about the apparition of time as an independent entity, 'on the progress of which the position of the pendulum depends' when *factually it is the motion of the pendulum, or any other equivalent, that we call the passage of time.* We will discuss this later, in great detail.

It is a subtle, but important point that Mach's criticism of the Newtonian concept of empty 'absolute space' as an existential basis for dynamics is applicable to the Newtonian concept of the 'absolute time' in empty space as well. The two cannot be separated because, while the notion of space is tightly linked to matter and its extension, the notion of time is linked to the evolution of the physical states of matter. Finally, as we shall learn, the evolution of matter and time itself are modified by the gravity of matter and energy in a universal way, as only gravity can do.

It is very important to realize that even today, we do not understand the origin of inertia and the commonplace and ubiquitous inertial forces. Therefore, the deep questions raised by the discussions by Newton and Mach remain unanswered even after Einstein's apparent success with his General Theory of Relativity. Isn't it surprising, and somewhat unbelievable, that the real force that presses us to the seat in an aircraft that is taking off, or the crazy forces we feel in a turn-twist roller coaster at an amusement park do not have a proper explanation in physics, even after centuries of refinements and advance in the theory of motion and dynamics? Which of the known forces of nature appear as the inertial forces? For me, it was a troubling surprise and a sure indication that our theories were incomplete. However, these are precisely the questions for which we will soon find definite and lasting answers.

1.2 Matter

Though the reason for physics is 'matter' and its 'dynamics', modern theoretical physics tries to treat space and time as more fundamental than matter. However, any rigorous approach shows that matter is the anchor for time. Time, in all cases imaginable, is the changes in the physical states of matter; time has no existence separate from matter. Time is localized with matter. *All of theoretical dynamics, where we assume a 'flowing time' to describe the motion in space, is a comparison of one motion of matter with another, with one of them abstracted as time.* Therefore, there is nothing physical called space-time, separate from matter and its physical states; space is where matter is, and time is contained within, and not outside space in a separate dimension. We will pursue this important observation further, in more detail.

Matter is the greatest puzzle faced by the 'matter with awareness', the 'conscious us'. The origin of space and time is of secondary importance to the origin of matter. The reason is that the only tangible reality that we sense is matter. The many qualia that we experience, based on which we form our intuitions of the external world, are all through the physical states of sensible matter. That is why scriptures of learned civilizations hypothesized on the origin of matter as prior to even the notion of God. The primary preoccupation and effort of physics have been always in understanding the ultimate structure and elementary units of matter, in the hope of understanding its presumably simple origins. There are two aspects of matter that are relevant at the core of our discussions. One is that all matter is source of gravity, without exception. This is very different from other interactions because there is a lot of bulk matter with no electromagnetic interaction – we label such matter as electrically 'neutral'. Only the matter with residual electric charge has the electromagnetic interaction. But, *all matter has gravity*. Matter mass is the charge of gravity. There is no 'gravitationally neutral' matter.

The second aspect of matter that is relevant for our discussion is the equivalence of mass and energy, discovered for the electromagnetic energy by Lorentz, and established in general by Einstein. Thus, energy is also matter and it is an equal source of gravity. This equivalence finally provided us a way of consistently speculating a possible origin of matter from 'nothing'. If the matter with positive energy is created with an equivalent amount of compensating negative energy of mutual interaction, the cherished notion of the conservation of energy and matter would be respected. Though we do not yet know about the nature of the origin of the Universe, a profound and useful relation emerges regarding its matter content, that it is equal to the total (negative) gravitational energy. Only gravitational energy needs to be considered in this large Universe, because cold matter is electrically neutral over large scales, and gravity is the only relevant long range interaction. Then, we expect that the positive rest energy of matter of mass m, $E = mc^2$ is equal to the negative energy of gravitational interaction, $E_g = -\sum Gm_i m/r_i$, where we have symbolically written the gravitational interaction of the mass m with every other 'piece' of matter (or energy) with mass m_i (or E_i/c^2) at the distance r_i. The interaction energy can be written as $E_g = m\Phi_u$, where Φ is the gravitational potential due to the entire matter in the Universe. For creation from nothing we need $E_g = -mc^2$, which is equivalent to stating that the cosmic gravitational potential at any point in the Universe is numerically equal to the square of the speed of light, $\Phi_u = -c^2$. The most surprising, and perhaps satisfying fact is that modern precision observations of our Universe seem to establish that this relation is maintained during the evolution of the Universe (Weinberg, 1977; Ryden, 2002). The Universe in which the total energy – the kinetic energy, the rest mass energy, and the gravitational potential energy – of matter is zero, is called a 'critical Universe'. The average density of matter and energy can be calculated to be about 10^{-29} g/cm^3. The directly and indirectly measured density of all matter, including dark matter, etc., is very close to this theoretical value.

This amazing relation, $\Phi_u = -c^2$, has profound implications to the nature of time, to the behaviour of clocks in motion, and to the theories of relativity in general.

1.3 Interactions

Our view of the relation between material bodies has evolved to a maturity and stability after the experimental methods and empirical evidence became the primary criterion for the acceptance of ideas about the physical world. It is quite amazing that the totality of all interactions among matter is more or less understood in terms of just four fundamental interactions, of which only two are effective over macroscopic distances. All the four interactions are intimately related to the possibility of life, and hence our existence.

The Earth is kept in an orbit around the Sun, the ultimate source of all the energy that sustains life, by the gravitational force. The same force also gives a structure to these globes, and holds both an atmosphere and us, close to the ground. The energy reaches us in the form of light and other electromagnetic radiation. The conversion of this energy to forms that we can use depends on a variety of complex electromagnetic interactions, in the cells of plants and in our own cells. The source of the Sun's energy is the fusion reactions in the core of the Sun, the mighty, but short range, strong nuclear interaction. This itself is facilitated through another short range interaction, called the weak interaction. This completes the four we know of, all of which are essential for us. Is it by design, or is it that life happened in tune with this prior environment of the fundamental interactions and the delicate position of the Earth relative to the Sun? Theologians and philosophers debate and discuss different views on this, but physicists just marvel at the precision with which the various factors seem fine-tuned to a near perfect harmony.

The importance of the interactions in understanding time is in the fact that they are behind the causal relations that establish order in the changes of the states of matter, both within a piece of matter as well as between matter at different parts of space. Time is the ordered sequence of changes in matter; the notion of time does not exist without both the changes of physical states of matter and the causal ordering of these changes.

Electromagnetism and gravity rule most of our daily life and activity. However, as already mentioned, gravity has the special role in that it is the only unshieldable universal interaction that affects, and is also sourced by, every kind of matter and energy. Electromagnetism has two kinds of charges, termed positive and negative, and magnetism is equivalent to the currents (movements) of these charges, which can also be of both senses. Therefore, both electric and magnetic fields can be completely nulled or shielded. Gravity has only one kind of charge – mass and energy of matter, and the interaction is always attractive and binding. Therefore, *gravitational fields cannot be shielded.* Gravity also has a magnetic nature, arising in the currents (movements) of matter. We use the common term, gravitomagnetic field, to describe it.

Since the currents of matter in space can be of both directions, gravitomagnetism can be attractive or repulsive. However, the gravitomagnetic effects are large enough to be observed only when there is dense and heavy matter moving with very high velocity, as in some extreme astrophysical situations. While the familiar attractive gravitational field is very important in the discussion of time and clocks, gravitomagnetic fields are only indirectly relevant for our focus on time.

Gravity is the only interaction that is known to affect the progress of time, or the rate of clocks. We already mentioned that motion affects the rate of clocks, as familiar in relativity. Gravity slows down clocks. This should not be confused with what happens in a pendulum clock in different gravitational fields. Gravity acts as a restoring force for the displaced pendulum in the pendulum clock. Therefore, the rate of such a clock does depend on the strength of the field. But, gravity is not relevant for the operational mechanism of a quartz clock or an atomic clock. However, all these types of clocks, irrespective of their physical mechanism, slow down equally and universally in the presence of matter and its gravitational potential. This gravitational effect is called the gravitational time dilation (Taylor and Wheeler, 1992).

There is one more fundamental aspect of gravity that needs to be mentioned. *Since energy is equivalent to matter and matter is the source of gravity, all forms of energy generate gravitational fields*. In particular, a gravitational field, which has energy, generates gravity! Gravity has its own gravity. This is the non-linear nature of gravity, unfamiliar in electromagnetism. This becomes important in the interactions of massive and very dense systems, as in common astrophysical situations of binary black holes. However, most of our discussion will be about time in the context of our familiar physical world, where such non-linear effects are too small to be seen.

1.4 Motion and Evolution

The simplest instance of a change in the physical state of matter is the change in its spatial being; a change in its position. This phenomenon of 'motion' is already a subtle issue, because a position can be identified only in relation with other material objects as reference. This is the relative space Newton alluded to. The coordinate systems that we take for granted cannot exist except as material references. This was recognized before the ideas of motion were formalized, and from time immemorial mankind has been using the reference frame of the 'distant stars' as the immovable reference to mark and measure motion precisely. The entire progress of astronomy depended on such a reference frame and the faith that it is more or less changeless.

Given such reference frames, the simple free motion of a particle represents an evolution that can be quantified (Fig. 1.2). A moment's thought will convince us that we have already discovered a clock and a marker of time, in a piece of moving matter. As Mach wrote, *any other motion or process can now be compared with the free motion of the particle, which serves as the 'time'*. Note the important content of the familiar mathematical expression for denoting the 'time evolution' of the simple motion of a piece of matter, $v(t) = \frac{dx}{dt}$; we usually understand dx as the increment in spatial distance and dt as the increment in time. *But, the dt is just another name for a dx, the incremental motion of another piece of matter that can be used as the reference for the motion of the former.* However, our linear motion clock (like C1 in the figure) is not convenient due to its implied boundless extent. Two ways out can be explored. One, to confine the uniform motion to a region by looping the path or reflecting the motion on itself (like C2, or like a pendulum), and two, to provide a fresh source of identical particles that move through shorter sections of space (as in an hourglass). All these methods have been explored in the quest of time-keeping in our history.

The term 'motion' is used generally in physics to indicate all physical processes that 'move', or evolve. Hence 'dynamics' is the general term that is used for any physical process in evolution. For example, the change in the temperature of a piece of hot coal is as much a dynamical evolution as that of a spinning top or a falling particle. Each of these is a legitimate clock, though only the spinning top would qualify as a practical clock for use, that too only for short durations. But, such physical phenomena are the only manifestations of time. Searching for time beyond the physical

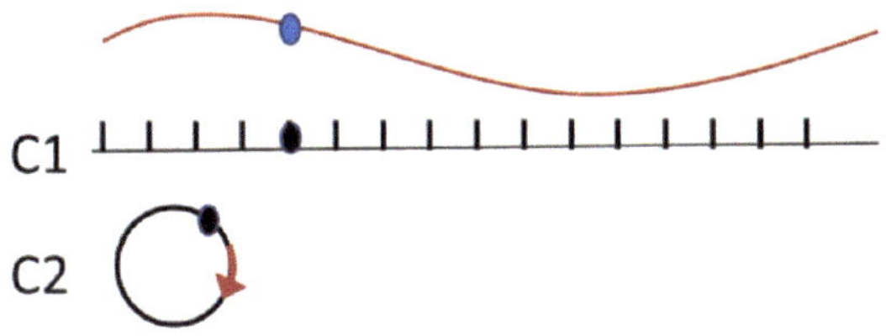

Figure 1.2 Motion serving as time and reference clocks. The regular motion of one bead (C1) serves as the time and the clock to measure the motion of another particle. A typical clock is a regular repetitive motion that is restricted in space (C2).

evolution of matter will turn out to be a stubborn exercise in utter vain, driven by a philosophical, or at best a psychological prejudice.

1.5 Time, Again

We are now ready to fix our ideas on time, in a manner suitable for a logically and physically mature discussion, while remaining rooted on a familiar conceptual ground. We go back and examine Newton's attempt again:

> Absolute, true, and mathematical time, of itself, and from its own nature flows equably without regard to anything external, and by another name is called duration: relative, apparent, and common time, is some sensible and external (whether accurate or unequable) measure of duration by the means of motion, which is commonly used instead of true time; such as an hour, a day, a month, a year.

We realize at once that his description of absolute time is not only operationally empty, as Newton himself realized, but it is also physically empty. It may be noted that while Newton tried to prove the reality of absolute space with a real experiment, he did not argue anything for the proof of absolute time. However, his notion of relative time is exactly what we always meant by time in our several thousand years long history. The evolution of modern science has not fundamentally changed our understanding

Figure 1.3 A representation and a realization of the physical time in a 'candle clock'. The states of the clock at two moments clearly show both the physical basis of time (in the evolution of the matter), as well as the non-existence of past. In this example, the 'future' is merely a piece of (remaining) matter, existing in the present, in space. Also, the candle clock and the conventional clock are interchangeable; they have equal status in representing time.

of time and its measurement. The measure of duration has always been by means of motion. But, the concept of motion used in the measurement of time has grown more general than the mechanical motion of objects in space, to the general evolution in the physical states of matter, denoted by the general term 'dynamics'.

Examining dynamics to understand the nature of time reveals an important conceptual point about 'time', first clearly stated by Aristotle. *Though we think of time as some continuously entity, existing through past, present, and future, a focused analysis makes it obvious that the past and present do not exist as a physical entity*! In this respect, time is very different from space (see Fig. 1.3). Just because we can represent time as an extended line with an arrow on paper or in our imagination, very similar to how we represent a spatial extension, we should not be misled to think of time has the same physical status. The way Aristotle put it is not only clear and precise, but also profound (Aristotle, 1991),

> The following considerations would make one suspect that it either does not exist at all or barely, and in the obscure way. One part of it has been and is not, while the other is going to be and is not yet. Yet time – both infinite time and any time you like to take – is made up of these. One would naturally suppose that what is made up of things which do not exist could have no share in reality.

As we will repeatedly mention, it is important to realize that our romantic notion of a continuously flowing stream of time, with the present sandwiched between the expansive past and future, is a psychological notion stored as memory in the brain, all existing together at present, only in the spatial confines of the conscious brain. *The immediate past, as a record in the local memory, is essential for detecting a change in the physical state of matter, by local comparison*. Physically, the past exists only as a record of simultaneities between physical events and reference clocks; this record exists 'now', in the present, stored in memory registers which are spatial elements. This is why both past and future have equal status in the memory storage; such spatial storage does not distinguish between past and future. How could it, for the record of time is a mere number – the number of cycles of a reference clock, which in turn is a suitable physical process, that has been counted while another piece of matter underwent changes in its physical states.

We will see at various occasions that the concept of a 'clock' embodies these concrete physical ideas about time, without carrying with it the irrelevant, imprecise, and inconsistent properties usually ascribed to time.

We will jump ahead a bit, with the reward of an important insight. Though we have not yet seen the relation between gravity and time, it is a fact that gravity affects time. However, gravity is an interaction. How can it affect something abstract and metaphysical? It cannot. Gravity is generated by matter and energy, and only by matter and energy. Space and time do not generate gravity. In fact, gravity is said to affect space and time only due to its effect on the spatial distance between material bodies and on the rate of material clocks. The assertion that an interaction that cannot be sourced by a physical entity can nevertheless affect the same entity, contains a potential inconsistency. In any case, to explore how gravity affects time we need to identify the physicality of time precisely.

This takes us to the notion of clocks. Ultimately, the notion of time is identical to the evolution of the physical state of a material clock. By asserting this, we are denying the separate existence of the abstract concept of time, from the physical construct of clocks. This goes back to the profound realization that there is no time without matter, in empty space. There is no time even with matter that has no evolution. Any amount of philosophizing cannot give life to a time separated from matter. We can prove this at once rigorously because it is an accepted fact among physicists and philosophers alike that time is affected by both gravity and motion. But, motion of what? Time does not move. Only matter and material clocks move. Therefore, the precise statement is that the motion of clocks affects their rate; not that motion affects time. Similarly, gravity is an interaction sourced by matter, which affects other matter. It is only by affecting the evolution of the physical state of matter that gravity affects the reckoning of time. Without material clocks in gravitational fields, all statements of gravity affecting time are physically meaningless.

However, just changes in the states of matter do not make time with a past, present, and future. There is one more concept that is inherent in time, and that is the notion of the accumulation of the quantity of durations, counted as the familiar and simple numbers. This endows time with a magnitude as well as an order. The notion of order is inherent because numbers are ordered. This is why we can rely on a periodic device rather than a monotonically changing device to reckon time. We are just counting the number of 'coincidences' or 'simultaneities' between a timing device (clock) and a physical process when we say that something is evolving in time. But the two – the clock and the physical process – are interchangeable, in principle. Time does not flow. It just accumulates, always remaining in its present. That is why one can conclude with logical confidence that past (and future) do not exist. There is only the 'now', and an accumulated number of past durations that exist in a localized spatial representation, as the only existence of physical time.

One may wonder whether any physical process could happen in an instant without extension. Just as a point in space without extension cannot be of use in rigorous and consistent physics, the notion of 'now' without extension does not seem physical. However, once we realize that *the evolution of physical processes in time is always a comparison between two physical processes, or their correlation, with one of them chosen as 'time'*, the comparison as a ratio makes proper physical and mathematical sense, all the way to the limit of zero extension. It is in this rigorous sense that 'now' without past and future is to be understood. However, there is another approximate sense in which this can be understood, without compromising on the conceptual integrity of the correct formulation of physical time. Since our contact with physical time involves a finite resolution δt, whether factual or imagined, 'now' can be assumed to have the smallest imaginable extension. What is to be noted is that, irrespective of allowing any amount of extension to our 'now', the past and future do not exist beyond the present!

These statements do not prohibit one from imprecisely talking about time itself as an abstract and independent entity, being affected by motion and gravity, as long as we understand that we are referring to the time evolution of the 'rest of the matter', after choosing the evolution of some reference matter as time. This kind of an unintentional separation is implicit in all our references to the notion of time. However, we should be clear about the underlying relations, so as to avoid physically meaningless speculations. This becomes important in discussions of ideas on time reversal, time travel, multiple 'streams' of times, etc., which we will briefly discuss later.

The reason why gravity cannot affect the past, or the future, is because these do not exist factually and physically. Gravity, or any other interaction for that matter, can only affect what physically exists. And that is only the present, the current physical processes.

1.6 Time and Space as Essential Unobservables in Physics

We must continue our scrutiny of space and time, to be absolutely sure that we understand the physicality of time and its inseparability from the physicality of matter. The fundamental physical theories that interpret and explain the behaviour of matter in nature are dependent on several unobservables and insensibles in their construction. Unobservables are postulated things with a well-defined mathematical and physical identity, but beyond our direct perception; their perception is indirect, only through observations on matter. In fact, it is not possible to give evidence for their existence, except as the manifestations of material existence and behaviour. *Entities like fields, wavefunctions, and even space and time are all unobservables of physics.* There is thus an obvious difference of degree and meaning between the reality associated with these unobservable theoretical entities and that of perceptible matter. The success of the physical theory is often taken as the evidence for the physical reality of such unobservables. While a rigorous natural philosophy will not be able to support or approve their reality with the same vigour and conviction as it might defend the reality of matter, there does not seem to be a way of avoiding such unobservables if we have to construct theories.

Space and time are the supreme and primary unobservables in physics, without any reality independent of matter, notwithstanding their modern status as the dynamical arena of all gravitational effects, as described in Einstein's General Theory of Relativity. However, they are essential for even starting to imagine about a physical theory, let alone construct it. Newton's Scholium on space and time in the Principia is a testimony for their essential need and role in building the physical theory. The difficulty faced by Newton in defining space and time in the construction of a physical theory without referring to matter has already been discussed. The situation is different from the notions of electromagnetic fields or the gravitational field because space and time are continua without material sources. Their existence and reality are the most primitively entrenched notions in our minds, and yet they are the primary examples of unobservables in physics.

It was the success of the Special Theory of Relativity that brought the unobservable empty space and vacuum as major physical entities into physics. It is in the human nature to provisionally accept many unobservable and insensible entities as plausible, and then believe in their physical reality, after prolonged familiarity. Both individual and sociological factors play a role in this transformation. For instance, most people (at least in India, but factually everywhere) are believers in several unobservable and unverifiable entities and causal effects, including occultic presences and astrological physical influences. For most physicists who grow up in such a society, there is no internal urge to question an unobservable that can be justified as useful to defend some aspect they have not understood. What is deemed irrational is finally decided by social circumstances, even in physics! The primary example is the widespread belief in the non-local and instantaneous influences that go across vast distances to affect quantum physical states of particles. Such an occultic influence is postulated only because physicists do not understand quantum correlations in terms of any cause-effect relation. But, there is no experimental evidence whatsoever for such an influence; experiments just examine the correlations between two independent measurements. Physicists are happy with, and even excited about, this spooky nonsense that goes against the fundamental premises of locality and causality in physical science, even though there is no theoretical or practical use in such a belief. (We will study the role of time and causality in this context, in Chapter 9.)

Space is some notion we can relate to as existing, with arbitrary extension (though in association with matter). Therefore, it makes sense to represent it by drawing three orthogonal lines and then locate objects and events in the three-dimensional space (Fig. 1.4). We are confident that the extended space and the material 'contents' exist as real, because our experience supports this stand. However, adding another dimension for time and drawing an extended line is not really legitimate. For every observer, only the instant of now exists; the extension of the time axis beyond the neighbourhood of the present, to the past or future, is not a verifiable physical fact by any means.

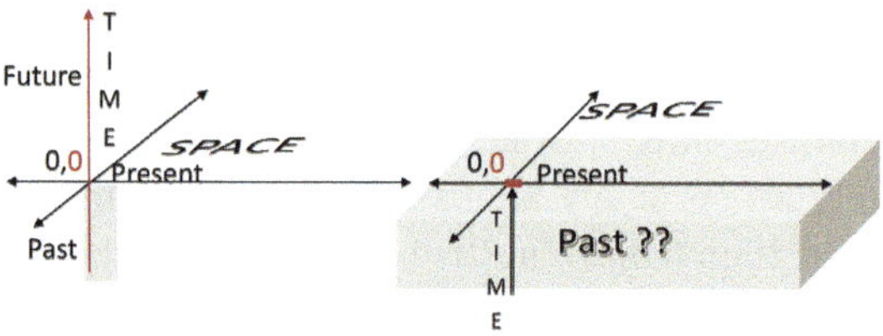

Figure 1.4 Left: Conventional view about the reality of space-time, representing 'time' as an orthogonal dimension to space, with both the past and future of the local present (labelled 0,0) in existence now. Right: However, since the physical time is inseparable from matter and its evolution, which exist only in the present, the conventional representation of time as an extended dimension is not legitimate.

Thus, even as a legitimate and 'official' unobservable of physics, the representation of physical time is faulty. The mathematical notion of time as another dimension similar to space, with infinite (or large) extent in both directions from the 'present', cannot be defended in physics, when closely scrutinized.

1.7 Clocks

Our familiarity with time is only through our familiarity with clocks. The most familiar material clock is, of course, our own life. As I will repeatedly assert, and give evidence for, time is not a physical entity that is separated from and independent of material clocks. Time is just the label we have ascribed to the collective notion of the evolving states of matter.

Just as marking and measuring space need a 'scale', which is necessarily matter-based, marking and measuring time require a matter-based device; that is what we call a clock. In space we measure distances, and from that we arrive at areas and volumes. In real terms, we measure not the distance in space, but the distance between material points (pieces of matter). Since space can be sensed in various ways, there is an enduring image and a concrete picture of the measure of space. Sensing and measuring time is not a straightforward task. In fact, we learn to read a clock in our childhood much later than we comfortably measure out spaces.

But, our everyday familiarity of measuring space will help us to understand better how we measure time. Then we will understand the nature of time better.

We measure space by superposing a material scale over space and counting coincidences. The scale has markings that are much shorter than the scale itself, enabling us to measure fine while measuring fast. We can measure space in any order of direction or partition. This is because we reason that the whole space under our scrutiny *exists*. We will see later that the properties of time are very different. Now we will just focus on marking and measuring time.

We can expect that measuring time needs a scale of time we can overlap on the 'time to be measured'. A better terminology would be 'duration' – we need a unit or scale of duration to measure durations of time. Then we have to just count the number of units in the duration to be measured. However, we immediately notice a problem. An extended duration does not exist! Unlike space with an extension, time does not have an extension. There is only the 'now'. Past is gone and future is not yet born. The only thing we can do, *and ever have done*, is to match the 'now' of our scale to the 'now' of the time we want to measure, with the limited resolution we have, and start counting. The count – a number – endures in space, in a memory, or on a piece of paper, or in the memory of a computer. When we say that one swing of a pendulum is a second, such a swing and the second do not exist. The 'now' of the pendulum might be just the beginning of the swing. Then counting the coincidences of the start of the swing with the events of any physical process, or counting their simultaneities, is the measurement of time. Duration exists only as number – the number of simultaneities of the events in a clock with the events of another physical process. And this number exists in some matter occupying space, and in the present. This is a precise statement about both the notion of time and its measure. Once this is understood, clear and convincing answers to several fundamental questions about time, some debated for centuries, emerge naturally.

We start with an apparently circular, but consistent definition of a clock, inseparably linking it with time. *A clock is what marks time and time is what is marked by a clock.* This underlines many features of what we call as time. This definition reminds us that time

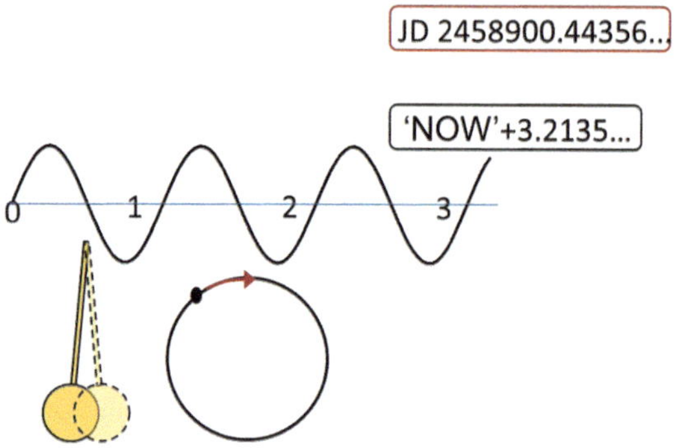

Figure 1.5 The truth about time, existing as the present 'now'. The physical existence of the past is factually in the present, as a mere number count of coincidences or simultaneities, recorded on a spatial register. The clock mechanisms can be varied, as a non-uniform pendulum or a uniform circular motion. It is the nearly uniform 'phase' (blue line) of the oscillator that represents time, not the actual motion. This is cumulative, irrespective of the direction of motion.

is nothing but the evolution of the states of matter, or dynamics in the generalized sense, of which motion in space is a special case. Not only that, the term 'mark' implies quantity and number. The elementary duration in a clock, whether it is repeated motion that is localized or uniform motion that is divisible in space, needs to be counted as a number to qualify the device as a marker of time (Fig. 1.5).

A patient and focused contemplation would then reveal that there is no time other than and beyond what is marked and counted by material clocks of motion. *Remove all matter, and then both clocks and time disappear.*

At once we see that a universal and standard time is a convention; it involves choosing a standard unit for marking time, agreeing on the actual numerical synchronization, and making sure that the different clocks with a variety of physical mechanisms and material details indicate the same durations between the same events. Already, we can realize that time-keeping could be a very complicated affair that calls for both technological sophistication as well as logical and scientific rigour. The factual situation that even identical clocks do not mark identical durations between common events, is a later realization that is at the heart of relativity and timekeeping. We have much to say about it. In fact, that fascinating and important topic is the focus of this book.

The slowly rotating Earth served as our primary clock for millennia, for both lay people and experts like astronomers. The earth-clock was a gift. It has a certain universality, apart from its remarkable stability. It was available to everybody. However, it needed the reference background of the fixed stars, or of the Sun's apparent position, that were not available always. Even when it was available, the expertise to read that clock accurately needed special training and scholarship. The earth-clock was not sufficiently 'local'.

We do not discuss here the fascinating history of the development of clocks, for which one may consult other sources (Landes, 2000; Jones, 2000). The most significant early advance, in terms of scientific discoveries, was the realization that the period of a simple pendulum depends only on its length and does not significantly depend on the amplitude of oscillations as long as the amplitude is not too large. The fact was apparently discovered by Galileo, much before the underlying physical law of gravity became known through Newton. In fact, the pendulum signified a double discovery involving gravity. One was that the period depended on the length of the pendulum and the local gravitational pull of the Earth, as $T = 2\pi\sqrt{l/g}$. The gravitational pull is more or less constant over the mighty Earth, changing only by about 0.5% over most places on the Earth. However, good pendulum clocks are stable enough to show these variations. The second one, a very important and fundamental discovery, was that the period does not depend on the mass of the pendulum – a bob of 100 kg on a wire has the same period as another pendulum of the same length made with a bob of 1 kg. This fact is the same natural miracle as the falling together of a tiny stone and a heavy lump of lead. This goes to the characteristic universal nature of the gravitational interaction, now called the (weak) equivalence principle. An oscillating bob of the pendulum is repeatedly falling and being pulled up, as long as its energy at the start is not dissipated in the mechanisms, or due to the drag in the surrounding air. Though the gravitational pull on the bob is proportional to its mass, the response of the bob to the force, or its acceleration, is inversely proportional to the mass. Thus, the mass drops out of the details of the motion. We will discuss this more formally later, because it needs to be understood well to build a complete awareness of the true presence of cosmic gravity and its effects, so crucial for the discussions in this book.

The controller and marker of time in the pendulum clock is the 'pendulum oscillator' of the clock. To make it mark seconds, the standard unit of time or its specific multiples, an elaborate mechanism for the 'conversion' of the units by a gear assembly is required; the natural period of the pendulum is converted to a unit second first and then to the larger multiples. The ingenuity in the development of pendulum designs and related control mechanisms resulted in good clocks of navigational accuracy and stability. The decisive contributions by Christian Huygens and John Harrison were particularly important. I should mention the remarkable fact that till about 1930, the best clocks available were pendulum clocks, with the Shortt-Synchronome pendulum clocks (Jones, 2000) with a master-slave design (by W. H. Shortt) reaching a stability better than 10^{-8}, losing only 1 s in 10 years! The 'master clock' was well isolated in a vacuum jar, while the 'slave clock' was chiming away a short distance apart, in the open, with both of them connected through electrical means, in a sophisticated feedback configuration.

Once the frequency of a periodic oscillator is fixed and known, the measure of time is the same as the 'phase' of this oscillator, each oscillation adding the phase 2π, or the familiar '360 degrees'. This is the convention because periodic phenomena go in circles, repeating on themselves after a fixed period, and a circle is 360 degrees. Whether the clock is the Earth itself or an atom, measuring time is keeping track of the accumulating phase of the oscillator.

The first quantum jump in the technology of the clocks happened from the discovery that certain materials like the quartz crystal generate electricity when stressed and they also extend and contract in response to an applied electric field. There are now many such 'piezo electric' materials, some natural and some synthesized, with the desired properties. An oscillating electric field results in the piezo mechanical vibrations of the crystal. Since mechanical vibrations have characteristic frequencies that depend on the material properties and dimensions of the accurately cut crystal, there are sharp resonant oscillations when the electric field

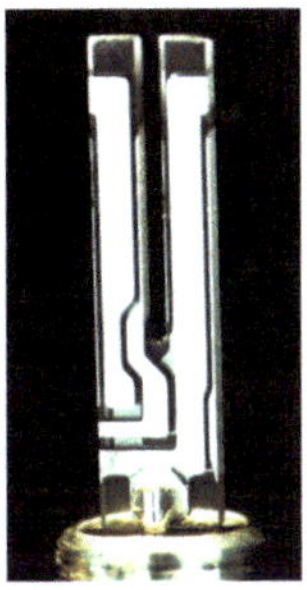

Figure 1.6 The millimeter sized quartz tuning fork with the natural frequency of 32786 Hz (2^{15} Hz), used in digital wrist watches. Then dividing by 2 repeatedly 15 times gives 1 Hz, or the period of 1 second. The stability and accuracy can be better than a part in a million.

frequency matches these mechanical characteristic frequencies. This defines the clock oscillator, where an oscillatory electrical signal is maintained stable by associating it with the piezo mechanical oscillations of the crystal. For quartz oscillators, the typical frequencies are in several mega-Hertz (MHz), determined by the typical crystal dimensions divided by the speed of sound in the material (Fig. 1.6). The stability of these resonances in carefully engineered crystals in controlled thermal environment can be better than a part in a billion, signifying a 'drift of time' amounting to only a second in 100 years.

The literal quantum jump happened when pristine quantum mechanical oscillators suitable for clocks were found in atoms. The energy levels of an electron in the atom, bound by the attractive electric force of the nucleus, depend on the particular element, but they are practically identical for all atoms of a specific element. The transition between such energy levels is a special type of quantum mechanical motion. A transition to a higher energy level can be forced by shining some light of a specific frequency, and a transition to a lower level happens with the emission of the light of the same specific frequency. The physical principle involved is simple, relating the difference in energy levels to the frequency ν of light, $E_2 - E_1 = \Delta E = h\nu$. The constant h is the Planck's constant. When this is done in controlled environments, the frequency is reproducible to a part in a trillion or better, defining the accuracy

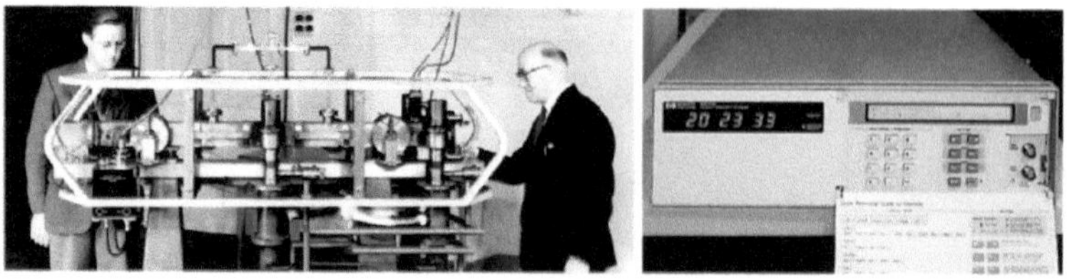

Figure 1.7 Left: The first Cs frequency standard constructed by L. Essen and G. Perry at NPL, UK. Right: A commercial HP Cs primary standard, used as the time-keeping standard in many time & frequency laboratories. GNSS installations like the GPS use these portable standards with a stability of a part in 10^{13}

and stability of the atomic clock oscillator. But the atom is not an oscillator, like the quartz crystal or the pendulum. Its energy levels are such that a transition takes place from one state to another only when the frequency of the radio (or optical) waves matches a specific value, given as $\Delta E / h$. Thus, the atomic clock is actually an atom-light clock. The real material oscillator in atomic clocks is only the quartz crystal oscillator used for synthesizing the high frequency radio waves (microwaves). The radio waves themselves also can be seen as equivalent to an oscillator.

The early Cesium atomic clocks used energy levels in the Cesium (Cs) atom, with relatively small difference, corresponding to the frequency of 10^{10} Hz (Fig. 1.7). The Cs atomic clocks have a stability of 1 part in 10^{13} to 10^{15}. What this means is that two such clocks keep identical time, with their difference not exceeding 1 second in 10^{13} s (300,000 years) to 10^{15} s (30 million years) on the average. The state-of-the-art atomic clocks use energy levels that absorb and emit in the visible optical frequencies, or even higher (10^{15} Hz). Consequently, these clocks are amazingly accurate and stable at a few parts in 10^{18}, projected to lose only a microsecond in 10000 years.

1.8 Dynamics and Time

Having seen the intimate and inseparable relations between time and the evolution of physical processes of matter, from simple motion in space to changes or 'inner motions' in the physical states of atoms, the plausible equivalence of time and material evolution should be more transparent. In fact, every dynamical evolution of matter can stand for the notion of time, with equal right. On the other hand, there is no support for a time independent of the physical states of matter. Examining the statements of the 'laws of dynamics' in physics will make this amply clear.

Consider the familiar Newton's law of motion, $m\vec{a} = \vec{F}$. In terms of the change of momentum the law reads, $d\vec{p}/dt = \vec{F}$. The rate of change of momentum is equal to the applied force. The denominator is the interval of 'time', but in all cases it is the rate of evolution of a clock – the motion of a material clock – which is assumed to be 'as uniform as possible' by the nature of its construction. Thus, dt can be a small change in the angle of the rotating Earth, an elementary vibration of a quartz crystal, or an elementary count of the frequency of light that is in resonance with an atom.

Another prominent law of dynamics is the Schrödinger equation that is applicable for the quantum dynamics in the microscopic atomic world, $i\hbar\partial\psi/\partial t = H\psi$. It specifies the temporal rate of change of a wavefunction, $\partial\psi/\partial t$, as driven by a function related to the energy of the system, called the 'Hamiltonian'. This is much like Newton's equation for dynamics, $dp/dt = F$. The time variable that appears here is identical to the one that appears in the Newton's law. However, if quantum mechanics is the true fundamental theory from which Newtonian classical mechanics emerges, then the seat of elemental time is quantum dynamics. After all, atomic clocks function within the laws of quantum dynamics. The inherent stochasticity and the quantum uncertainty are made negligible in the construction of such clocks, using certain amplification mechanisms. But one can see the nature of time clearer in the Schrödinger equation. This is because the wavefunction ψ represents the totality of the physical state of matter; the infinitesimal change of the parameter called time, represented as ∂t, is similar to the infinitesimal change in the physical state $\partial\psi$. One could symbolically write $\partial\psi/\partial\zeta$ where ζ is another wavefunction. Then, we are comparing the change of state of one material system to the change of state of another. Therefore, both the numerator and denominator in these laws of dynamics represent material physical systems that evolve dynamically. Only, the one that appears in the denominator is very carefully chosen to be of a uniform and universal nature, as far as we can know, to qualify as 'time'.

We conventionally take a state with a regular periodic behaviour that can be replicated in copies of the state, and call it a standard clock, and a standard of time. Let us now briefly examine how matter that evolves according to quantum mechanics is important for the notion and the definition of time, and for the construction of reliable clocks.

1.9 Time, Frequency, and Quanta

The fundamental idea of quantum mechanics is that matter, or particles, have 'some' wave property. In the early days of the theory, it was thought that the particles behave also as waves; this double role of matter was called the wave-particle duality. But nobody could figure out what property of matter was 'oscillating'. Also, while certain limited class of experiments could be consistent with a literal wave property of particles like electrons or neutrons, most experimental situations suggest that it is not the particle itself which is behaving like a wave; the particle does not split in two paths and recombine, etc. The severe foundational problems that arose a century ago in the context of the wave-particle duality were solved only recently, by recognizing that the theory of quantum mechanics is not in fact the fundamental theory. Quantum mechanics and its 'wavefunction' represent the mechanics and evolution of the statistical ensemble, or a collection, of dynamical histories (Unnikrishnan, 2018b). The new description of quantum dynamics needed an amazing completion of Hamilton's mechanics, which W. R. Hamilton introduced in 1832–35 (Hamilton, 1833, 1835). However, we do not need to discuss those details for dealing with our topic of time and gravity. We can go ahead assuming that some entity with a periodic property is operative in the dynamics of microscopic matter like atoms.

The periodicity in quantum mechanics is associated with the fundamental dynamical quantities of energy E and momentum p. The connecting relations are $E = h\nu$ and $p = h/\lambda$, where ν

is the 'frequency' and λ is the (de Broglie) 'wavelength', both being quantities usually associated with a wave. These are the same relations as the ones obeyed by the waves of light, or electromagnetic radiation. An atom has various discrete set of energy levels. When the electron in the atom changes its physical state from one level to another, it emits or absorbs a fixed quantum of energy, corresponding to the difference in the energy of the two levels, $E_q = E_2 - E_1$. Since an atom is bound by the electromagnetic force, the quanta are photons, or light, with the energy $E_q = h\nu$. Light in free space also obeys the relation $\lambda = c/\nu$. Therefore, light emitted by the atoms can provide a standard of time or frequency (number of oscillations per second) and a standard of length (the wavelength is the length of a complete wave of light).

There is one fundamental problem though. The energy levels (rather, their difference, which is the only quantity that is observable) are not precisely defined – there is the small fundamental 'Heisenberg uncertainty' in them. If the electron is excited to the higher level, it spontaneously jumps back after some average 'lifetime' ΔT. Then, the difference in energy is uncertain by $\Delta E \simeq \hbar/\Delta T$. Therefore, the frequency of the photon is also uncertain by $\Delta\nu = \Delta E/h = 1/\Delta T$. However, the longer the lifetime of the excited state, the lesser is the uncertainty in the frequency of the light that excites the atom, as well as in the frequency of the light that the atom emits when it de-excites. It is clear that a good atomic clock needs energy levels that are long-lived, with large decay times and small uncertainty in the frequency. To summarize, the matter waves hypothesized by L. de Broglie remain invisible in any quantum process, but the energy of the electron in the atom is quantized in discrete energy levels. The difference in the energy of two levels is equal to the energy of the photons emitted or absorbed by the atom, during the transition of its state between those energy levels. Making sure that the electromagnetic radiation (light) has its frequency and energy exactly equal to the difference of the particular energy levels in the atom, by maximizing the absorption of the light by the atom, 'locks' the frequency of the radiation to the stable fixed energy level difference of the atom. That is how the atom acts as the stable reference for the frequency of light. Light can be easily detected and its frequency is accessible and measurable accurately by standard

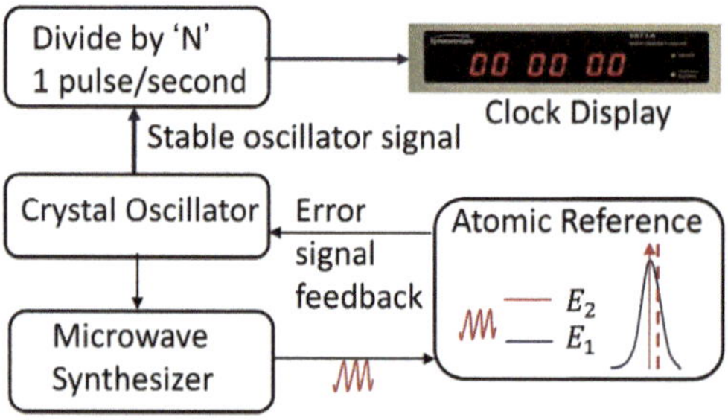

Figure 1.8 Scheme for the generation of the 'standard second' referenced to an atomic transition. The crystal oscillator is used as the reference to synthesize the electromagnetic radiation that drives the atomic transition. Any deviation of the frequency of the radiation from the exact resonance causes less absorption and an error signal, which is fed back to the oscillator to correct its frequency. The stable output frequency is counted and divided suitably to get the '1 pulse/second' clock signal.

techniques. Then, counting a fixed number of the oscillations of light provides a rock steady and universally reproducible 'atomic second' (Fig. 1.8) – that is is the modern standard of duration. (Jones, 2000; McCarthy and Seidelmann, 2009)

1.10 Being and Time

I touched up on the factual point that the extended time, from the far past to the distant future and passing through the 'now', does not exist in the same sense as the extended space 'exists'. When we say that the extended space exists, what we really mean is the generalization of our awareness that the different material existences in our 'present', including that of our friends and acquaintances, are in different spatial locations. The notion of the present or 'now' is identical to that of 'being', and it contains the idea of simultaneity.

The meaning and the relation of the two concepts, being and time, have been discussed in the philosophical thoughts from ancient times. In fact, the puzzle of the transition from nothing to being has been the perennial theme of 'the origin', in physics, philosophy and theology (Heidegger, 1927).

There is an important difference between the philosopher and the physicist in dealing with the fundamental notions like time. A philosopher is free to define the notion, with certain demonstrations of consistency with common sense and intuition. But the physicist doesn't have that kind of freedom; he cannot stray far from what is empirically known, or can plausibly be known. Since 'measurability' is always associated with the physicist's notions, he cannot *decree* the nature of the fundamental concepts; his freedom is limited to consistent interpretations.

Another important fact about the evolution of concepts in the society is relevant. When a concept changes its character in the physicist's view in a significant way, it affects all of physics; the change is tangible not only in the scientific papers, texts and applications of technology, but also in the public life of the society that uses many technologies. An example is relativity's time. But, when the philosopher makes a comment on the nature of these very concepts, however influential it might be, there is no immediate change noticeable in the way the society uses these concepts; there is not even an urgency to understand the new view, let alone assimilate it. Hence, philosophical debates are contained, whereas scientific debates can percolate far. (After all, there are no philosophers of time in the committees for defining and maintaining standards of time to be used by all, globally!)

Often, physics provides a precise framework for the formulation of a working definition, meaning, and interpretation of concepts that are important for philosophy. This is particularly so when the most basic physical concepts like time and space are involved. Therefore, ancient philosophers, who were also the natural scholars who contemplated on the physical world, perhaps found it easier to define the concepts and then formulate the questions and plausible answers. The link between physics and philosophers (not philosophy!) is very fragile and tenuous in modern times. Thus, these philosophers do not any more have the advantages those of a past era had. A philosopher is qualified to analyse and comment on the physicist's theoretical construction of the physical world when the philosopher is able to understand the logic and meaning behind the physical arguments. Similarly, physicists would engage in a philosophical debate when philosophy has a critique that can endanger the integrity of the fundamental structure of physics. This possibility of interaction slowly diminished in the 20th century, and by the second half, the debates have been confined to the superficial. In this book, we have to confine to a discussion that is mostly within physics. But one can glimpse the possibility of providing reliable and precise input regarding the same concepts in the philosophical sphere.

It is possible to define and go forward with some precision, with the concepts of Being and Time, starting with the intuition from the physical world as seen by physics.

Being is the physical state of an entity, and time is then the evolution of this physical state. 'Being now' does not have a duration in itself. Or, if it has, then such a duration is smaller than anything we sense and distinguish as duration. A stack of numerous ordered states of being is existence and that has duration. Thus, time is the evolution of being. Time is not external to the physical state of being. It is within a *dynamic being*. A reminder to the missing physical factor in the Zeno's paradox, of the arrow that does not move, may be a helpful analogy. Zeno argued that at an instant of time the arrow is stationary at a region equal to itself in space. One cannot get motion from a sequence of stationary states. The resolution involves recognizing the dynamic quality of motion – that of the velocity or momentum. Being is not a stationary physical state, because it is a state of matter, consisting of many particles and their interactions within and outside. The state of being is subject to the dynamics of change. And this change is what we experience and understand as time. As mentioned already, the change of state can be a simple positional change (as of an arrow or falling stone) or a complex process like the ageing of a living body. Since the notion of physical states include also the mental states, all varieties of time naturally are included in this formulation. It is precisely because of this relation between time and being that we have the possibility of the multitude of times in relativity. If time were entirely separate from being, then time would have been an absolute single entity, independent of the physical state of matter. Only matter can move and undergo changes by interacting with other matter; both result in the modifications of time. But, abstract time or space cannot move or interact with matter.

Out of all physical entities that undergo changes of physical states, one gains priority because of its gravitational dominance over every other, and that is the physical system of the entire Universe. Its state of being is the only one that can affect and control time significantly, through its enormous gravity. This then holds its own time, related to the evolution of its being, and provides us with the notion of master clock, more or less the same everywhere in the vast Universe. We will discuss in detail later this important and profound feature, not realized when our fundamental theories were developed and completed.

In physics, it is common and popular to depict physical events conveniently in a space-time diagram, with one axis representing the extension in space and the other (conventionally vertical) axis representing the progress in time. In such a space-time diagram, just 'being' at a fixed location is equivalent to uniform 'motion' along the time axis. Since space-time diagrams treat time as a coordinate akin to spatial coordinates, one keeps moving 'forward' in time, whether or not one is moving in space. That is, one cannot but move in time continuously. However, it is important to note that this representation has hidden assumptions, of a real reference clock that is not specified, but assumed to be existing at that location. Without assuming a material entity with a progressive change of its physical state, there is no way to reckon time and mark it on an axis, even in a conceptual diagram. Therefore, there is a significant difference between 'being' and 'being with a clock'. The latter has two distinct clocks, one the dynamic self and another, a material clock as the reference.

In Heidegger's influential view, 'time' is implied as primordial to 'being' (Heidegger, 1927). In contrast, the reasoned view maintained consistently in *Gravity's Time* is the opposite, with 'being' rooted in matter as primordial to 'time'.

1.11 Astronomy's Time

The archetype of time, with all its attributes known across civilizations, derives from the relative motions of the celestial bodies and the myriads of natural phenomena that originate from the peculiarities of these motions and relative orientations. Its most notable feature is the apparent periodicity, juxtaposed on the non-periodicity of an individual's life. The steady rotation of the Earth remains among the best clocks known to humankind (Jones, 2000; McCarthy and Seidelmann, 2009). Only in recent times, with the development of the atomic clocks that are more stable than the steady rotation of our planet, we have come across the necessity to shift our dependence to these devices as our reference clocks. Isn't it an amazing fact in itself that till about 1960 the standard of international time was defined and distributed based on the rotation of this giant rotating ball in space, carrying its mountains and oceans, its atmosphere, and the billions of its inhabitants, most of whom are not even aware of this imperceptible rotation?

The revolution of the Earth around the Sun, the self-rotation of the Earth, and the intermediate periodicity provided by the lunar motion around the Earth completed our early time-reckoning devices. Understanding these motions and their relations to all other motions of other visible planets, all in the background of the 'fixed' reference background of identifiable stars, shaped our logic, mathematics, and science. Astronomy's time and human time are so intimately linked that we cannot separate them at any level. The development of various calendars and sky charts and the handling of their details even by common astrologers bear testimony to this inseparable link. The only distinction one can possibly make is regarding the formal aspects of precision, which requires sophistication in observations, data handling and model building.

In a sense, perhaps superficially in the full context of our discussion of time, astronomical time is gravity's time. The existence of life itself is dependent on the special gravitational bond between the Earth and the Sun. Besides, the whole process of the evolution of life to the present form critically depended on the gravitational environment of the Earth, among many other factors. However, this is not the sense in which we will discuss time as 'gravity's time'. We want to explore the universal physical aspects of the relation between time and gravity, going beyond the interesting details of time and its varied facets in relation to our life on Earth.

1.12 Our Earth Clock

One way of realizing the importance of the favourable chances for our existence, that arose in the particular configurations in the solar system of celestial bodies, is to imagine a counter-factual situation and its consequences. What if the rotation of the Earth was much slower, perhaps taking the duration equivalent of several weeks? One can speculate that the evolution of life would have been different, but how different is difficult to quantify without detailed studies. Various biological processes and 'rhythms' would likely be different. But, it would also be likely that many things might not be entirely different. After all, humans live in places where the difference between day and night is small, for extended durations. In any case, what is clear is that the self rotation of the Earth with a period much shorter than a year, with only about a degree of movement in its orbit around the Sun, helped us considerably in reckoning time in our lives, with high precision. The common phrase '360 degrees' goes well with this celestial arrangement, *with the 'degree' signifying a day's worth movement of the Earth relative to the fixed stars*. In fact, I think that this peculiar unit of the angle, dividing circle by 360, came directly from the cycle of the relative motion of the distant stars during the year of about 360 days, noticed from antiquity. Between the day and the year, the lone 'Lune' was a bonus, with its well-defined phases and its pleasing soothing appearance.

The Earth takes about 86400 seconds for a full rotation. Or, rather, the second is defined as 1/86400 of one rotation of the Earth, measured as the passage of the position of the Sun in the sky. What is important is the stability of this rotation, owing to the large mass and the relatively fast rotation (angular momentum) of the Earth. Despite its atmosphere and ever changing topography, the perturbations of this rotation are not readily noticeable. Yet, the day is not an unchanging 86400 seconds worth forever. The long-term change in the length of the day is estimated to be between 2 and 3 milliseconds per century. This is not a smooth and steady change; however, when the fluctuations are averaged out, there seems to be a long-term steady change. The Earth itself is about 4.5 billion years old – about a third of the age of the Universe. The day is longer by about 25 seconds in a million years. Hence, there was a larger number of days in a year, as much as 400 days or more, hundreds of million years ago. The most amazing fact is that there is direct evidence for this in some coral growth records, much like the periodic growth marks in tree trunks (Wells, 1963).

The figure 2.5 ms/century amounts to a drift of less than a part in a trillion (7.9×10^{-13}) in the rate of the earth-clock, which is better than a rudimentary atomic clock. However, there are larger short-term fluctuations in the earth-clock, of a fraction of a millisecond, on yearly and even weekly scales, mostly owing to the atmospheric perturbations. Therefore, the Earth is not a very stable practical reference clock, stable enough for modern applications on human time scales. The state-of-the-art atomic clocks, however, are not far from reaching a stability of a part in a million-trillion (10^{-18}). Therefore, we now depend on certain atomic clocks for our stable time, against which we can easily measure and monitor the variations in the earth-clock. Since we cannot readjust the Earth's rotation physically, we must keep track of the deviations, occasionally adding a 'leap second' to the standard time based on the atomic clocks, 'by order', to maintain the synchronization between the atomic clocks and the earth-clock.

In the influential and comprehensive text *Elements of Natural Philosophy* by W. Thompson (Lord Kelvin) and P. G. Tait published in 1879 (second edition) (Thompson and Tait, 1879), a future superior clock that excels the earth-clock is described,

> The ultimate standard of accurate chronometry must (if the human race live on the earth for a few million years) be founded on the physical properties of some body of more constant character than the earth for instance, a carefully-arranged metallic spring, hermetically sealed in an exhausted glass vessel. The time of vibration of such a spring would be necessarily more constant from day to day than that of the balance-spring of the best possible chronometer, disturbed as this is by the train of mechanism with which it is connected: and it would certainly be more constant from age to age than the time of rotation of the earth, retarded as it now is by tidal resistance to an extent that becomes very sensible in 2000 years; and cooling and shrinking to an extent that

> must produce a very considerable effect on its time-keeping in fifty million years.

The same text goes further in anticipating later developments, this time ascribing the idea to J. C. Maxwell,

> The period of vibration of a piece of quartz crystal of specified shape and size and at a stated temperature (a tuning-fork, or bar, as one of the bars of glass used in the 'musical glasses') gives us a unit of time which is constant through all space and all time, and independent of the earth... But this depends essentially on one particular piece of matter, ... insuperable practical difficulties which are experienced when we attempt to make exact copies of them... The recent discoveries due to the Kinetic theory of gases and to Spectrum analysis (especially when it is applied to the light of the heavenly bodies) indicate to us *natural standard* of matter such as atoms of hydrogen, or sodium, ready made in infinite numbers, all absolutely alike in every physical property. The time of vibration of a sodium particle corresponding to any one of its modes of vibration, is known to be absolutely independent of its position in the universe, and it will probably remain the same so long as the particle itself exists. The wavelength for that particular ray, i.e. the space through which light is propagated in vacuo during the time of one complete vibration of this period, gives a perfectly invariable unit of length; and it is possible that at some not very distant day the mass of such a sodium particle may be employed as a natural standard for the remaining fundamental unit. This, the latest improvement made upon our original suggestion of a Perennial Spring, is due to Clerk Maxwell.

All self-rotating celestial bodies can serve the role of a precision clock. We use only the Earth's rotation as a precision clock at present. However, pulsars in our galaxy, which are rotating neutron stars that are about as massive as the Sun, are more stable as clocks owing to their larger mass and faster rotation, as well as due to their isolated nature. Some pulsars match the long-term stability of advanced atomic clocks. An ensemble of such pulsars will be required for reliable time-keeping because there are occasional unpredictable changes, albeit small, in their rotation rate. There are recent global astronomy projects that aim to use the network of such pulsars to operate a time standard network, usable for navigation and other applications that require high precision timing with long-term stability.

To acquire good intuition and understanding of the standard of time based on the Earth's motion, the excellent video source https://youtu.be/IJhgZBn-LHg may be consulted.

1.13 The Moon Clock

The motion of the Moon has been used in all civilizations to mark time, and the lunar calendar is as important as the solar calendar. In fact, the rapid motion of the Moon in its orbit, compared to the slower movement of other easily observable bodies in the solar system, led to the Moon playing the most prominent role in the definition of a relatively recent standard of time called the 'ephemeris time'. This gained some acceptance for use in precision astronomy, and as the standard time during 1956–66, before the universal acceptance of the atomic time, because the ephemeris time is defined in terms of the accurate periodicity of the gravitational orbits of several celestial bodies. The argument was that the standard of time based on the precision of a fundamental physical law like Newton's inverse-square gravity, will be precise, stable, and reliable. However, as one can imagine, the ephemeris time is not easily accessible for direct observation and verification, and requires astronomical instruments and techniques as well as precise calculations using the details of the law of gravitation and dynamics.

The history of near-space astronomy teaches us an important lesson regarding time, time-keeping, and clocks. Centuries of data on the Moon's position relative to the Earth and the stars had the evidence that the Moon was speeding up in its orbit, as measured by the earth-clock. This fact persisted even after the tidal correction that actually slows down the Moon, but raises its orbit, is applied to the Moon. It is the peculiar (counter-intuitive) property of gravity that if a body in orbit loses kinetic energy and slows down, its orbital radius increases, rather than shrinks. This is because the body is kept in the orbit by the balanced action of gravity and the centrifugal force. A decrease in the velocity, because of the loss in the kinetic energy due to the tidal friction, implies reduced centrifugal force that can be balanced only by weaker gravity, which requires orbiting at a greater distance. In any case, even after applying all the identified corrections, the Moon seemed to speed up in its orbit.

The puzzle was solved when engineered clocks that were better than the earth-clock became available. The real reason turned out to be the 'slowing down' of the standard reference clock used to measure the Moon's motion, which was the earth-clock! Its rate of rotation had retarded over the centuries (again, due to tides). And for us Earth dwellers, who used the Earth as the standard clock, all phenomena seemed to have speeded up a bit over the centuries! *The important lesson is that we cannot distinguish between the changes in the rate of physical phenomena from the changes in the rate of the standard clocks we use.* The flow of time is identical to the dynamical evolution of the 'clock', which is the physical process that is used directly or indirectly as the reference process. The flow of time is assumed to be regular, but its regularity is only as good as that of the reference physical process, which in this case was the rotation of the Earth. Just like measurable space is defined by the relative position, and the inertial motion of one body against a reference of other bodies, measurable time is defined relative to a reference clock, explicit or otherwise. Any change in the reference process affects the rate of the 'flow of time', which in turn gives the impression that the time evolution of the physical process being observed accelerates (or decelerates), because we assume that the flow of time is uniform, due to our belief in the reality of a pre-existing physical time.

The incident of the mistaken judgment of the speeding up the orbital motion of the Moon teaches us some fundamental lessons about time. There is no time other than what is measured by a clock, and the regularity of the passage of time is only as good as the regularity of the reference clock.

1.14 Ephemeris Clock

The ephemeris clock is based on the trust in the law of gravitation, which makes the planets and their satellites go in periodic orbits

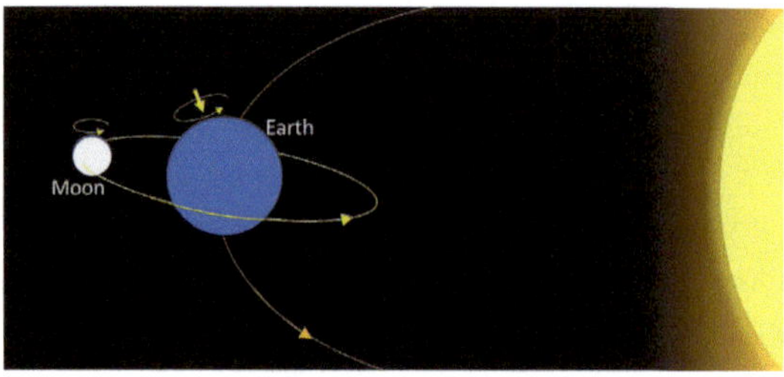

Figure 1.9 The standard of time based on the gravitational orbits of celestial bodies is more stable than those based on their self-rotation. This is the justification for the ephemeris time.

in empty space (Fig. 1.9). They are not circular orbits, but the periodicity of the full orbit is more stable than the rotation of a mass like the Earth. Newton's law of gravity applied to the multi-planetary system is the basis of the ephemeris times. (Small corrections described by Einstein's General Theory of Relativity can be incorporated.) But calculating the planetary positions taking into account the gravitational action of every significant body in the solar system is a formidable task, made possible by developments in both mathematical techniques and the methods of numerical calculations, aided by computers.

A feel for the remoteness of such a 'scientific' definition of time from what is easily accessible for common use can be indicated by comparing the definitions. If we had a stable pendulum of a standard length, one might define the 'second' as equal to 'one complete oscillation of the pendulum'. If a pendulum provided, for some reason, such standard seconds everywhere, then it would serve as the reference clock. But, the pendulum is not universal because its period of oscillation is affected by many factors. The earth-clock, on the other hand, is much more stable. The second is defined as 1/86400 of the time it takes to complete one rotation. This is a more reproducible standard of the 'second'. However, it involves more expertise in observations. The ephemeris second is defined based on the periodic orbital motion of the Earth around the Sun relative to the distant stars, which takes a year. It was defined (in 1960) as 1/31556925.9747 of the tropical year for 1900. (The year needed to be specified for accuracy, because there are small variations in the orbital duration. The full definition is more complicated because it involves Newcomb's theory and tables of the motion of the Earth around the Sun, for the calculation of the orbits, or the ephemeris.)

The trend changed, from the definition based on the long period standards (much longer than the second) to the very short period standards, with the new generation clocks of the 20th century. When the atomic time became available, the ephemeris standard was quickly replaced with the atomic second in the System International (SI), defined in 1967 as the duration of 9,192,631,770 periods (cycles) of the radiation corresponding to the transition between two hyperfine levels of the ground state of the Cesium-133 atom (isotope with the atomic weight of 133 units). The particular number of cycles in the definition arises from matching the atomic second to the ephemeris second at the time of the definition (the earth-clock second is slightly longer than the ephemeris second, requiring 9,192,631,937 cycles of the Cesium hyperfine transition).

Another comment may be relevant, though obvious. A reference clock like the rotating Earth or the orbit of a planet is a single physical specimen. Therefore, its stability and 'reproducibility' can be checked only by comparing the successive cycles against some other standard, if available. In sharp contrast, any number of atomic clocks can be constructed, all of which have the same reference transition frequency within a small range of variations (which is dictated by fundamental quantum physics). So, they can be compared against each other for their stability. In terms of the general principle, the atomic time resembles the ephemeris time, being decided solely by the fundamental physical laws. While gravitation and classical dynamics are the basis for the ephemeris time, electromagnetism and quantum dynamics are the basis for the atomic time. Loosely speaking, one can say that both are based on 'orbits' of fundamental interactions.

1.15 Cosmological Time

The visible Universe around us and its apparent vastness prompted many fundamental questions from time immemorial. While the understanding of the immediate neighbourhood of the solar system developed significantly through progressively refined observations, modelling, and the theoretical advances, the larger physical Universe in the cosmology of the early 20th century was not very different from the Universe that was familiar to the humans of early civilizations. That Universe consisted of the vast number of distant stars that were distributed over the whole sky, the star-dense region of the Milky Way (which was later identified as a vast 'galaxy of stars' in which the solar system was embedded), and a few small fuzzy nebulae. Astronomy with telescopes was confined to mostly the solar system and the Milky Way, even as late as 1900. There were no galaxies other than the Milky Way in the scheme of our Universe before the 20th century.

The distant stars, more or less 'fixed' in their positions in the sky, were the stable markers for the motion in space. They provided a reliable navigational reference. They also enabled marking time with the regular motions of the celestial bodies, including the rotation of the Earth.

This elementary picture saw a drastic revision after 1920, primarily due to the installation of large telescopes of refined optical quality (with apertures larger than 1 m), aided by advances in photography and spectroscopy. It was recognized that there were numerous dense systems of stars, or galaxies, separate from the Milky Way. Study of the 'extra-galactic' nebulae progressed fast. After Henrietta Swan Leavitt established the relation between the period and luminosity of the 'Cepheid variable stars', astronomers could estimate the distance to other galaxies using the Cepheid stars as the 'standard candles'. Spectroscopy made it possible to determine the radial velocity of stellar systems, by measuring the Doppler shift of specific spectral lines. In 1928, the decisive discovery by Edwin Hubble of the systematic recession of distant galaxies, with their velocity proportional to their distance, revealed an evolving and expanding Universe. Thus, in a brief period of 10 years, and well after the revolutionary theoretical developments in dynamics and relativity, our knowledge about the Universe, its matter content, and its evolution changed drastically. Suddenly, the Universe was much bigger, and contained vastly more matter in innumerable galaxies distributed in every direction. And the Universe seemed to have had a beginning, when all the matter in recession was together, about 10 billion years in the distant past.

The steady expansion of the Universe, manifest in the systematic recession of the galaxies, is an evolving physical process that defines a giant clock and a cosmological time. Since it is thought that the recession of the galaxies is identical everywhere in the Universe, the cosmic time is the same everywhere in the Universe. One can say that the cosmic clock is synchronized everywhere. We will discuss the cosmological clock and the universal cosmic time in detail, in Chapter 4.

1.16 Anthropic Time

Any reasonable theory of life and its relation to the physical world cannot deny that all animals and even plants have a 'sense' of time. This sense is not necessarily neurological. What is clear is that the sense and awareness of time in humans are the result of the long evolution of life through various stages. However, time is a very special and intimate notion for us humans. One reason for this is our necessary preoccupation with our memories and projections, aided by our remarkable capability to imagine our being in the past as well as in the future, in arbitrary situations created in the same imagination. Hence, our *imagined existence* is more or less free in both space and time, with arbitrary motions allowed back and forth in all dimensions. We cannot dwell on a discussion or even speculate about its purpose, use, or basis. However, the anthropic time in general has more possibilities than the physical time. Our sense of time is formed by millions of years of biological evolution, immersed in the cycles of physical time determined by the motions of the Earth in the solar system. We have imbibed and inherited the impression of the regular flow of time, most likely from the regularities in our terrestrial habitat.

It is by contrasting the physical time with the anthropic psychological time that we learn about the firm basis of the physical time on matter, and the many constraints it must respect in the physical world. The most 'disappointing' property of the physical time for us is its unidirectional nature and its strict continuity, without arbitrary jumps. Our fantasies, in mythology and in science fiction alike, are rich with time travels and time projections. Seeing the future is the ultimate skill that is still very much in vogue, with a majority of the populations believing in such a faculty, by varying amounts! Trying to see and read the future is perhaps our way of forgetting our memories and the past, hoping ever for the better. In any case our notions of expectations of time are fashioned and influenced strongly by our ability to retain memories of events in the past and their manipulations in our imagination, to create the desired future using elements of the past.

1.16.1 Memory, History, and the Psychological Time

With the human abilities of language and speech, both the individual memory and the collective memories store past events, providing a record of the lived time. The psychological time with its many subjective aspects cannot be the basis for the study of the physical time. However, our very reasons for taking time so seriously as a topic of study and a notion to be understood in depth are rooted in the questions we explored in our psyche during our extended evolution. These questions are not just confined to purely physical aspects of time. Rather, the dominant concerns are psychological as well as physical, apart from the obvious practical aspects of daily logistics of life. The structure of the brain and its ability to retain and retrieve memories have critical roles in our concepts of time. It is explicitly mentioned here lest we forget this, as well as the fact that the physical states of neuronal matter underlie the entire edifice.

1.16.2 Cinema and Time

Cinema illustrates best the relation between Being and Time. In fact, the core realization behind the invention of cinema and its precursors is the correct relation between being and time. The 'frame' illustrates 'being' and the scroll in motion, 'time'. There, time becomes synonymous with movement and life. But, only just so, because the experience of time and motion is psychological and psycho-neural. Further, cinema illustrates what notions of time are consistent with the evolved common sense and the common psychological time, and what manipulations of time clash with the common sense. It is only in cinema we have reached our full possibility of imagined worlds, sampling every desire and every possibility in space and time.

While the spatial possibilities are vast in cinema, time is the supreme element that gives cinema its unique character that cannot be obtained anywhere else. One example is that of visual time reversal.

Fast or slow moving time is not contrary to common sense, because it is familiar to the psychological time and our routine imagination. 'Time travel', in the sense of going back and forth in the memory of real or imagined events, also is not considered as unusual. However, the visual depiction of the time reversal of certain physical events by reversing the sequences of motion do clash with our common sense and experience. Do the fundamental principles of physics allow the possibility of time reversal? Hans Reichenbach wrote in his widely read monograph *Direction of Time* (Reichenbach, 1971),

> Negative time is very well illustrated by the use of motion picture films reeled in reverse. Processes which in this way of reeling appear compatible with physical laws are reversible; those which appear incompatible are irreversible. Mechanical processes like the throwing of balls and the lifting of a vase from a table appear quite natural in the reversed movie. In contrast, the burning of cigarettes, the pouring of cream into coffee, the breaking of pottery, and so on, lead to incredible happenings in reversed time. Such occurrences represent irreversible processes.

We will briefly address the important question of time reversal in a later chapter.

There is a valuable lesson we can learn from the relation between frames and time in cinema. Each instant of time is a frame in space, on the reel of the film. The past and the future of any chosen 'now' in the reel are all such frames, existing only as records in space, in proximity, in the same reel. In the digital era, the past and future of now are just a set of numbers with their existence only in the present, in a spatial record of memory registers. We will see that this lesson contains some deep truths about the nature of time.

The 'local' nature of time, encapsulated in finite space, with its reality only in the present, can be seen explicitly in the case of cinema. A reel of film, or a pen-drive of digitally encoded sequence of frames, with the items ordered in space, all present 'at once' simultaneously, is the entire and accurate representation of time.

1.16.3 Animals and Time

The animal sense of time is very similar to the human sense of time, in its primitive relevance. This does not mean that the human interest in time is similar; admittedly its scope is enormously larger. However, it is worth noting that there is no abstract and philosophical worries about the notion of time among animals, though one would expect that they have a sense of the finitude of life. The animal sense of time is presumably limited to the comparison of their own life with that of others, in the background of their regular and slowly changing environments. Sudden changes of external physical markers that indicate time perturb animals enormously. The physical time is more real than the psychological time for animals, the more so as we go earlier in evolution. The physical time is simpler to analyse than all other notions of time. As I stressed a few times already, the physical time, in one form or the other, is identical to the evolution of the physical states of matter. Animals seem to see just exactly that physical nature of time, manifest in the changes in their physical habitat.

As witnesses of lived durations, plants and trees are more ancient and long-living than animals. However, in the absence of a nervous system and memory, there does not seem to be a way of associating a *sense of time* to plants. Ironically, the two physical factors that are genetically tightly fixed in plant life, especially during the growth phase, are time and gravity.

Are humans the only species that worry about the brevity of their natural life spans? Is it at least partially because of their awareness of the deep conceptual extensions of the familiar time, like eternity?

1.16.4 Clock as Life and Life as Clock

This is a brief aside into the territory of philosophical debates, exploring what kind of reliable input can be provided to philosophy from the physical theory. Essentially, we are looking at the relation between clock time and lived time. Conventionally, the notion of time carries with it the property of continuity and the character of 'flow', with us always positioned in the present. These are properties ascribed to the physical time from our experience of the psychological time. But, these properties of time are much debated in both ancient and modern philosophy. Surprisingly, the physical time needs only the property of continuity, which is already an assumed property of the evolution of physical states. After all, formal physics describes the physical evolution of matter with the aid of differential equations of dynamics. Continuity is an essential notion in this description. Can the human notions of the past, present and future be made more precise, with a clear distinction between the past and the future? This is formally easy to do for each individual in the physical world, as we will see. The difficulty arises when we want to compare the present, past and future of different individuals. By 'individual', we mean here a general notion of a single reference point (localized region) in space with a local clock to mark time. We can also consider a world around us consisting of points that are stationary relative to us, though we need to realize that we can know only our infinitesimal neighbourhood in space, and the time in our local clock. Knowing beyond this needs *interconnecting* signals. It is precisely the physical considerations about knowing beyond the immediate neighbourhood that make the theories of relativity.

The difference between past and future arises with such contrast in the human notion of time because, and entirely because, of the fact of the impossibility of time reversal. There is no rigid notion of such two-worldliness in space because we are free to move between spaces – left or right, up or down, forward or backward. But, we cannot go backward in time. The much advertised equivalence of time to space, as just another 'dimension', is factually a created illusion. It is just a mathematical construct, with hidden supplementary structures, and without any mapping to the experienced notions of time. Along with the possibility of time reversal comes, naturally and surely, the ability to freeze time and make everything eternal. The physical time allows neither.

Since it is an important argument, let me state it in some detail. Suppose we assume, in alignment with the conventional views on the physical time, that time flows in a direction, and that it is possible in principle to go back in time (time reversal). Now, if we are moving in space, reversing the direction of motion will allow us to retrace our steps. However, this could be done at any speed. In particular, we can be at zero speed, static at a location in space. In fact, we have to stop an instant, to reverse the motion in space. If one can change the direction of one's current 'movement in time' along the axis of time, then it implies that one can also slow down the progress of time arbitrarily; then being static in time is a special case. In fact, the point of being static is inevitable in time reversal. But, being static in time means the end of time evolution.

1.17 Time in Close-Up

We are now ready to consolidate the fundamental and cardinal facts about time. In reality, the time that flows, with the past, present, and future as its parts, does not exist! It is our extreme familiarity with matter and space with extension that created our beliefs about time, and an analysis in close-up shatters these beliefs. Since the physical time exists only as the elementary changes in the physical states of matter, time is a local entity in space, just as matter is. In particular, the past and future are not part of the physical time – these are physically non-existent, except in the present, as spatial records. There is only 'now', locally, at our own spatial location. Every material location has its own 'now', and we can connect them through interactions if we please. Then, there are physical restrictions on which 'nows' are perceptible (or observable) simultaneously. The talk about a flowing time, its present straddled by an infinite (or very long) past and future is illusive, existing only in our imagination. *The past is merely a present record of elapsed durations, stored and updated in a spatial location*, like the brain, or a book, or a computer. A careful and patient analysis will show that time is localized in space, indeed in matter, as a physical process in progress now, and as a number that denotes the count of accumulated coincidences or simultaneities with a reference process called time (clock), which also occupies a proximate spatial location. For brevity and future use, we will denote this number as the time-number. It is identical to what we call the 'past', but it exists only as a number, localized in the spatial neighbourhood.

The fact that the past does not exist, except as memory in matter and mind, is the real reason that there cannot be any time travel. We are completely misled about a physical time with an extension into the past and future by that extended line we draw in a space-time diagram, giving us the false impression that the past as well as the future exist as real physical entities. This false impression is caused by our endowing time with the familiar characteristics of space.

The immediate consequence of this extension-less nature of time is that there is no sense in talking about retracing or modifying a time evolution that has been already reduced to the time-number. Just as we cannot go to a spatial location that does not exist, there is no way to go to a point of time that does not physically exist (see the evaporated candle-clock in Fig. 1.3). Since time at some location is in the totality of the local physical processes, and not outside, only the present exists as real. Reversing a physical process does not reverse time, to note a simple fact. A physical process in the opposite order is also counted in *increments*, and the time-number keeps increasing. The past accumulates as the localized number, ever increasing (Fig. 1.10).

Since this is a very important point, let us examine it in some more detail. I have an accurate and stable reference clock that can indicate the finest time scale I need. The working of the clock is a physical process, which is repetitive (for convenience) and countable. Each event that happens as a physical process in my neighbourhood has a well-defined coincidence with a count of the reference clock – a simultaneity of one process with another. The process is in the matter and not separate from it. Therefore, it does not exist 'in time' anymore as the 'past', except as the particular number that denotes its coincidence with the reference clock. The entire material physical system, including the reference clock, evolves in its dynamics, accumulating simultaneities and time-numbers. For example, the event 'my parents got married' was a

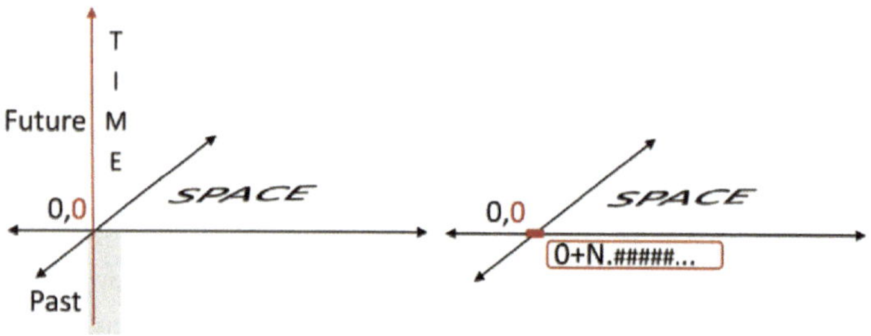

Figure 1.10 The fact about physical time. Left: The conventional view of space and time in modern physics, as 'four-dimensional space-time' with extensions in both space and time directions. Right: The factual situation, in which neither the past nor the future exist physically. The entire past at a spatial location is encapsulated in mere number counts of durations, existing in matter occupying space, and not existing in time. The 'time-number' (denoted as *N.##...*) that exists as a record locally in the present is the only physical existence of the past.

physical event with a time-number and a location. The location exists now and it existed then. But, there is no physical process or its time existing now – they are denoted by their identifying numbers at some spatial locations.

We can count the simultaneities by opening our fingers, and after reaching ten, folding them in the reverse order, but the counts of simultaneities increase only in one direction. *Reversing the order of motion has no implication on the progress of time.* The direction of elementary physical time is thus obvious. Every physical process increases the count because it is just the number of elementary physical processes. There is no negative counts of processes, events, etc.

Since the 'past' (or the future) does not exist as a point in time to go to, time reversal or time travels do not make physical sense. Those notions are matched to the flawed intuitions we picked from thinking about time as equivalent to space, with extensions beyond the neighbourhood.

And finally, I stress the physicality of time, its inseparability from matter. This is established by at least one undeniable proof – that time is affected by motion and gravity, as we will see in detail. *When we say that time is affected by motion, it is implied that the entity that is affected is capable of motion*. Space and time cannot move, only material clocks can move! Hence, time has a reality only as the evolution of the physical states of matter. Time is inseparable from matter. This important insight about time cannot be overemphasized.

1.18 Summary

In this introductory chapter, we glanced over several general aspects of time and occasionally dwelled deeper to highlight certain crucial ideas. We asserted the physical nature of time, firmly and inseparably linked to matter and its physical states. I argued that the changes in the physical states are the only manifestation of time in the physical world, without exception, and that empty space without matter and its dynamics has no time. We explored the nature and function of clocks, and briefly explored the major types of stable clocks humankind used through the development of civilizations. The special and permanent role of the rotating Earth as a clock was highlighted. Clocks and time, being the evolution of the physical states of matter, are subject to modifications due to the fundamental physical interactions. While clocks are constructed to be immune to such perturbations, the universal interaction of gravitation cannot be shielded. The gravitational modification of the rate of clocks is one of the most profound physical effects on time. There is also the mysterious effect of time dilation of any clock that is moving relatively, without any physical interaction.

Clocks are affected by both motion and gravity. But physicists are used to saying that time is affected by motion and gravity. Since only material clocks can be in motion, and be in a gravitational field, the correct statement is obviously that clocks are affected by motion and gravity. Then, after identifying time with what clocks mark and measure, as we have been insisting for reasons of logical and physical rigour, we can talk about how time is modified by motion and gravity. Historically, what was realized first was the modification of the rate of clocks due to motion – the time dilation of relativity. The equally profound and non-intuitive gravitational time dilation was discovered by Einstein, well before he completed his theory of gravity, which came to be known as the General Theory of Relativity. This was because the theory was based on extending the Special Theory of Relativity (STR) by assuming that the results of the STR are applicable in accelerated or general non-inertial motion as well. We will first explore the features of gravity as an interaction, to learn how it affects time, irrespective of the nature and mechanism of the clocks, living or non-living.

Chapter 2

Gravity

Until Einstein's General Theory of Relativity (GTR) was formulated, gravity had nothing to do with the passage of time. The only connection one could make was at the level of actual devices to mark time, because gravity was needed for pendulum clocks, and even for water clocks and hourglasses to work. Also, behind all astronomical clocks, there is the gravitational interaction as the agent responsible for the regular orbits and their periodicity. However, there was no inkling that gravity can slow down the rate of clocks, albeit minutely. Before the full theory was completed in 1915, Einstein had derived the correct gravitational time dilation of clocks in a gravitational field, using only the equivalence principle and the motional Doppler shift of light (Einstein, 1911). One may wonder what the spectral shift of light has to do with time. Since the argument is simple, we can discuss it right away.

Einstein's key assumption was the local equivalence of being in a uniform gravitational field $\vec{g}$ and being in a reference frame (a room) that is accelerating at the acceleration $\vec{a} = -\vec{g}$, for all physical processes. The two situations are indistinguishable locally. This is called the Einstein Equivalence Principle (EEP), which is a generalization of the Weak Equivalence Principle (WEP) that states

Gravity's Time
C. S. Unnikrishnan

ISBN 978-981-4877-72-5 (Hardcover), 978-1-003-25656-4 (eBook)
www.jennystanford.com

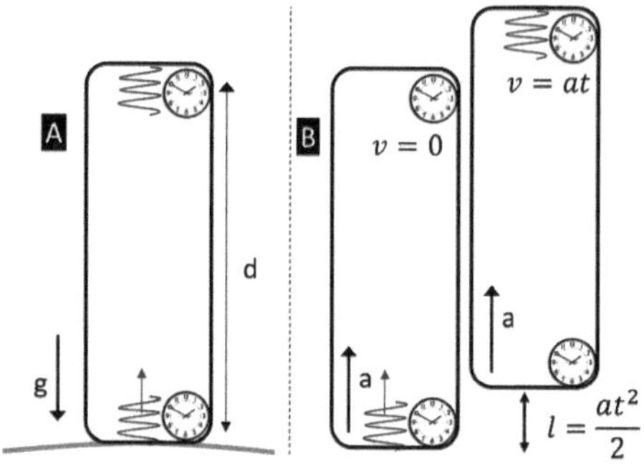

Figure 2.1 Einstein's derivation of gravitational time dilation. (A) Pulse of light propagating up in a gravitational field g. The frequency of light at reception is measured. (B) Similar experiment in a room that is moving up at the acceleration a, in empty space. The two frames denote the start at the velocity $v = 0$ and after a delay t, at velocity $v = at$. Light is received Doppler shifted due to the changing velocity, towards the red of the spectrum. From the assumed 'equivalence' of A and B, it is expected that the frequency of light is redshifted in the gravitational field (A) as well.

the equivalence of inertia and mass (we will discuss these principles in greater detail soon).

The physical situation is indicated in Fig. 2.1. Consider an accelerating laboratory in which a pulse of light of frequency f (and wavelength $\lambda = c/f$) is emitted near the floor, upwards, in the direction of motion at time $t = t_e$. The pulse is detected at a point which is at a distance d, at time $t = t_d$. However, the velocity of the detector at t_d is different from that of the emitter at time t_e because of the acceleration. During the interval $\Delta t = t_d - t_e$ the room and its contents have changed their velocity by $\Delta v = a\Delta t$. Therefore, the pulse must be detected Doppler shifted,

$$f' = f(1 - \Delta v/c) \tag{2.1}$$

The fractional difference in frequency is

$$(f'' - f)/f = \Delta f/f = -\Delta v/c) = -a\Delta t/c \simeq -ad/c^2 \tag{2.2}$$

The last step that used $\Delta t = d/c$ is an approximation for δt small enough to ignore the higher-order terms. For an observer inside the accelerated room, the entire room is at rest. However, he will feel a force proportional to $\vec{a}$, akin to gravity, in his rest frame. So, it is now easy to argue from the EEP that light in a gravitational field should be shifted identically in frequency, which was a profound discovery Einstein made from this simple argument. This was verified in a very famous experiment by R. V. Pound and G. A. Rebka Jr., nearly 50 years later (Pound and Rebka, 1960). However, the cream of Einstein's genius was not just this discovery of the gravitational redshift.

After noticing that the frequency of light changes in a gravitational field, Einstein made a characteristic leap of interpretation with solid logic; he argued that a frequency, which is the number of oscillations n per unit time T, could change at the measurement end only if the rate of the receiver's clock changed relative to the clock at the emitter! In the accelerating room, applying the STR, the speed of light has the universal value c. Frequency is the number (of oscillations) in a unit of time. *A number (count of oscillations) is an invariant and couldn't change from emission to reception inside a room at rest. Therefore, what has changed must be the unit of time itself.* The clocks at the points of emission and detection must be running at different rates. If such a phenomenon happens in an accelerated frame, it must occur in a gravitational field as well, according to Einstein's Equivalence Principle. Since the pseudo-gravity in the rest frame is equivalent to a real gravitational field, two clocks separated by the height d in a real gravitational field would have a relative time dilation given by

$$\Delta T/T = (T_2 - T_1)/T \simeq gd/c^2 \tag{2.3}$$

with the clock at a higher point (farther from the source of gravity) running faster (recording more elapsed time). Clocks would slow down when gravity is increased.

Ever since, the gravitational time dilation is a part of physics, and it was confirmed directly on clocks only about 60 years after the prediction. Compensating for this phenomenon is essential in navigational systems like the GPS, as we will discuss in detail later. (Clocks in a gravitational field indeed run slower, as many experiments show. The behaviour of clocks in an accelerating frame is not known, however, because such an experiment has never been done. The subtle reason will become clear when we discuss the modification of time in motion and in accelerated frames, in

Chapter 3.) The gravitational time dilation is not the same effect as the gravitational redshift of light, though the two are related. Clocks and time enter only indirectly in the description of the gravitational redshift. In contrast, the gravitational time dilation is seen by directly comparing the durations marked by two precision clocks. I note that the clocks need not be in different gravitational fields for the effect; the force field (the gradient of the gravitational potential) $g = -d\phi/dx$ can be the same, but the gravitational potentials ϕ are different at the locations of the clocks. This is a crucial detail of physics that will turn out to be very important for our discussion of the gravitational time dilation of clocks due to the gravity of the matter in the Universe.

What exactly is the property of gravity by which the interaction affects universally any and all clocks, independent of the actual mechanism of the clock? The GTR does not address this physical problem; the theory just asserts, in a mathematically precise way, that the rate of clocks, and hence time itself, is affected in the presence of masses. To understand Time we need to understand Gravity in more detail.

2.1 A Slow Thought Experiment

Equipped with the essential knowledge that matter and its gravity slow down the rate of all clocks, we can do a relaxed thought experiment on the cosmic scale. Imagine empty space without matter (other than the experimenter) as our entire Universe, much like the space of Special Relativity and the space assumed in current theories of physics. Empty space is the same everywhere and in every direction; it is homogeneous and isotropic in all situations. What is 'time' in such a world and how do we measure it? We need some well-defined state of matter which can evolve, to give us a physical clock and a notion of time. This could be a particle going around in a fixed trajectory, or an oscillator like a pendulum. Let us assume that such clocks are also available. This is the world of Einstein's theories. Now we want to focus on gravity's effect on our clocks. Let us slowly introduce matter into the previously empty Universe, keeping the constraints of approximate homogeneity and isotropy. Matter has gravity and as matter is added, our clocks should respond with the gravitational time dilation. The gravitational potential of ϕ translates to a time dilation factor ϕ/c^2. Therefore, if we denote the rate of a clock as ν_0 ('ticks per second') before any matter was added, the modified smaller rate is $\nu_\phi = \nu_0(1 - \phi/c^2)$. We can keep on adding matter, and eventually the approximate expression for the time dilation will not be applicable, when the factor ϕ/c^2 approaches 1; then the time dilation becomes very large. The eventual result would be that if the total potential is large enough, the rate of a clock is significantly altered by the matter in the Universe. We will see, with a calculation, that *the factual gravitational potential due to all the matter in the Universe is very large, nearly a billion times the gravitational potential of the Earth*. In fact, it is indeed large enough to make $\phi/c^2 \simeq 1$. However, such an obvious influence is not even mentioned in today's physics! Not much was known about the amount and the distribution of matter in the Universe when Einstein's theories were completed. Time in the factual matter-filled Universe is certainly very different from the time in Einstein's theories in empty space. What we experience as time is what could be called a 'renormalized' time, dilated and regulated by the enormous gravity of the smooth (homogeneous and isotropic) matter distribution of the Universe.

Therefore, it is very clear that the time displayed by any kind of clock is the time *already* gravitationally slowed down significantly by all the matter in the Universe. We have to reverse the execution of our thought experiment, and remove the vast amounts of matter from the Universe, to transform the factual Universe to the empty space of Einstein's relativity theories. *As the matter content of the Universe and its gravity are reduced, the gravitational time dilation decreases; the rates of all clocks increase and eventually the possibility of time disappears with the last bit of matter.* The only logical possibility is to conclude that the notion of time ceases to exist without matter. So, time does not make any physical sense in an empty space without any matter or energy. This implies the strong conclusion that time does not exist without gravity.

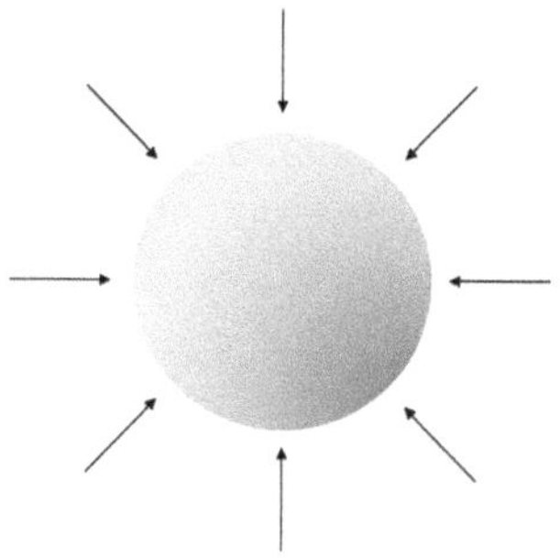

Figure 2.2 Gravity as an attractive force, sourced by a mass.

2.2 Gravity as a Force

Gravity is most familiar to us as a pulling force (Fig, 2.2). Almost every living organism is sensitive to this. The quantitative and systematic studies on the rate of motion (free fall) under gravity's pull were taken up by Galileo. The force is large, because the Earth is large; hence, for the practicality of the studies, Galileo had to slow down the motion using inclined planes. (In comparison, the forces that we generate in our bodies are far stronger, as it is clear from the tiny sizes of our bodies compared to the Earth! This is because the muscular forces are electromagnetic in nature, which is incomparably stronger than gravity.) A simple pendulum is another way to study the force of gravity and it offers a simple way of determining the local gravitational field, accurate to a part in thousand. This is possible because of the most important feature of gravity that was discovered by Galileo – that the gravitational acceleration, or the rate of increase in the speed of motion under gravity, does not depend on the mass or other properties of the falling object (Galileo, 1638). This is called the Universality of Free Fall (UFF). When a bob tied to a string is released from an angle, it is really falling under gravity. The constraint of the string pulls it up again, and then the motion repeats, as the oscillations of a pendulum. The Universality of Free Fall implies that the period of the pendulum depends only on its length and gravity, and not on the mass or the nature of its bob.

The law of the force of gravity was discovered and formulated mathematically by Newton (Newton, 1846). As everyone knows, the gravitational force between two particles or regular spherical objects with masses m_1 and m_2 at the separation r is

$$\vec{F}_g = -\hat{r}G\frac{m_1 m_2}{r^2} \tag{2.4}$$

where G is the gravitational constant that represents the strength of the interaction. $\hat{r}$ is the unit vector along the separation r between the bodies. The negative sign indicates attraction.

Newton had formalized another universal fundamental law before discovering the law of gravity – the law of dynamics that describes all kinds of motion when any kind of force is applied on a body. Newton's (second) law of motion states the proportionality between the force and the acceleration of the body,

$$\vec{F} = m\vec{a} \tag{2.5}$$

This is more suggestively written as a relation between the effect and the cause, as $\vec{a} = \vec{F}/m$. The force F can be of any kind, including gravity. The quantity m is the mass of the body. In the context of this law, the mass is also called *inertia*, or the inertial mass, as the resistance offered by the body to the attempt to move by a force.

We face a deep feature of physics right here, still remaining as an open problem, even after the many insightful discoveries by luminaries, like Newton and Einstein. What could be mysterious and remaining as unsolved in these two simple laws? Well, these two laws are completely independent, a priori. One deals with a specific force of nature (gravity) and the other is a general law for dynamics, applicable to all bodies and all kinds of forces. It so happens that the same person, Newton, discovered both.

Note that a quantity with the same physical nature – mass – appears in both the laws! It looks reasonable that gravity is proportional to the masses, just as the electric force is proportional to the electric charges. So, mass is the 'charge of gravity'; that is, it would be accurate to call the mass appearing in the law of gravity as the gravitational charge. However, what appears in the law of dynamics that defines the response of a body to an applied force is apparently the same quantity, the mass of a body. *Why should the mass do the double role in the physical stage as the charge of gravity and also as inertia*? Why is the charge of gravity, and not the charge of electromagnetism, acting as inertia? That is most amazing and puzzling.

However, if they are indeed the same quantity, for some reason, the universality of free fall that impressed Galileo follows. Taking one body as the Earth, we can write the gravitational force as

$$\vec{F}_g = -\hat{r}G\frac{Mm}{r^2} = \vec{g}m \tag{2.6}$$

where $\vec{g}$ is the Earth's gravitational field. When this field pulls on the body of mass m, it accelerates and the law of dynamics gives the acceleration,

$$\vec{a} = \vec{F}_g/m = \frac{\vec{g}m}{m} = \vec{g} \tag{2.7}$$

which does not depend on the mass! From a heavy stone to a feather or an elementary particle, all will fall with the same acceleration; the same amount of space is covered in the same duration. What a remarkable result! To highlight the fact that the true and ultimate reason for this universality is not understood, one must distinguish the masses in the two laws by designating them with different names, as the 'gravitational mass' and the 'inertial mass'.

2.3 Gravity beyond the Pulling Force

Imagine that we are at the middle of two large bodies, as large as the Moon or even the Earth. Now, the pull of gravity is from both sides, equally strong. A small object at the middle will stay motionless and will not fall towards either body; the net force is zero. That does not mean that there is no gravity at the middle! There is twice the matter around and there must be some measure of gravity that indicates this fact (Fig. 2.3A). This measure cannot be the pulling force, because that cancels exactly at the midpoint. In fact, there are two other familiar measures associated with gravity, both of which are as important as its pulling property.

One measure is as visible and familiar as the force of attraction. That is the tidal force, familiar from the spectacular ocean tides. If we imagine a small extended body, as all bodies are, at the midpoint of the two masses in the earlier example, all parts of the body except the central point will get pulled by each of the masses. On one side, the pull from the mass on that side will be slightly larger, owing to

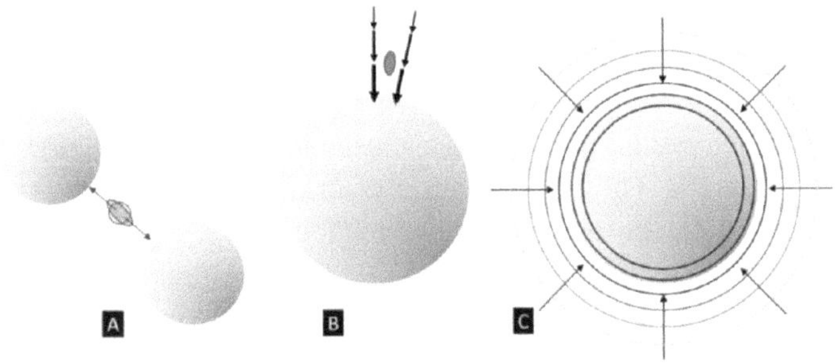

Figure 2.3 (A) The tidal force due to the difference in the gravitational forces near the midpoint of two massive objects. The typical effect is the shape distortion of a small spherical body. (B) The tidal force near a single massive body. The spherical test-body is radially elongated, and two falling objects (along the arrows) come closer. (C) The fundamental measure of gravity – its interaction potential energy – is insensible by direct means. The gravitational potential is constant along each circle (sphere), but radially reduces progressively as GM/r.

the shorter distance, than the pull from the mass on the other side, which is at a slightly larger distance. The result is that the small body gets stretched out and distorts. This is the tidal force. Even if there is only one large mass, this happens because the gravitational pull is more at a nearby point than at a farther point of the extended body. The tidal force is the *difference* in gravity's pulling forces on any extended body (Fig. 2.3B). While this is very important in all bulk gravitational phenomena and in the mathematical formulation of the GTR, the spatial tidal forces do not figure in our discussion of the gravitational modification of time.

The other important measure of gravity is subtle, but real. To understand it correctly, and easily, let us first look at a large metal plate connected to one terminal (say, +) of a battery cell marked 9 V. We would expect that the plate is at the 'potential' of +9 V (designated as V_0), relative to some ground at 0 V. The whole plate is at the same 9 V. If we imagine mapping the electric potential from the surface of the plate to the ground at some distance d meters away (which may be another plate connected to the other terminal), the potential varies from 9 V to 0 V, over the distance d m. Then we say that the 'electric field' E is V_0/d V/m. Any charged object, like a small charged piece of plastic, will get pulled by this field (or some time pushed, since the charge can be of both signs). The product $E \times q$, of the field E and the amount of the charge q on the small body, is the pulling force in this case. If we increase the distance d to the ground, the pulling force reduces, because the potential difference from 9 V to 0 V is now over a larger distance. The gradient (variation in space) of the potential is the pulling force, called the electric field. But *the electric potential has to be as physically real as the electric field, because one cannot have a gradient of a quantity unless the quantity itself is physically real.* The important point here is that just because we can detect only the 'variations in the electric potential' (the electric field, for short) directly, and not the primary quantity of the potential itself, we should not make the mistake of denying the physicality of the electric potential.

If we fold the plate into a more or less closed shape, the plate is still at the same potential; the *inside* of the closed shape does not see the ground at 0V, now outside. In the closed interior, moving from one location to another does not change the potential! Then the inside of the closed plate is an 'equipotential' space. There is no electric field there and a charge will be free to move without any electric force. However, the potential is not zero; it is still the same 9 V at all locations inside. So, the third measure of interaction – the potential – is really subtle. Its presence is detectable through a force on a charge only when it varies in space. This variation, or the gradient dV/dr is the field. While the gradient of the potential is zero inside the closed plate in our example, it is non-zero and felt as a force outside. We see that the (interaction) potential is the fundamental physical quantity. It is proportional to the quantity of charges present and defines the energy of interaction.

The potential can be felt also if it is not constant in time. A time-varying potential is felt as a force by moving charges. This is easy to see because 'motion' is equivalent to a change in the spatial location. If the potential changes during the motion, we have a situation of a variation of the potential in space, as far as the charge is concerned, though it is indirectly induced by the motion. The charge then feels the time varying potential as a force.

There is another very important aspect of the potential. It is the quantity that defines the electric energy of a charged body; *the product of the charge and the potential at its location is the energy of interaction*. We do not sense the energy directly; we sense only the changes in the energy, as the forces. However, that does not mean that energy is absent when it is not sensed. For example, the 'temperature' of a body is a measure of the energy of the random thermal motion of the molecules in the body. But we can sense only the *difference* in temperatures between the body and a 'sensor'. In quantum mechanics, the dynamical evolution of the physical system is entirely described in terms of the energy of the physical state, and not in terms of the forces. A system with more energy evolves more rapidly. Yet, what is measurable is only the difference in the rates of evolution of two states. Since a device like the atomic clock uses the energy levels of atoms, we finally see the importance of this discussion for our topic. An atom with more energy is equivalent to a more rapid oscillator. We will come back to this discussion later.

The situation regarding gravity is identical to the one in electricity in many aspects, except the fact that gravity is always attractive, owing to the fact that its source, the mass, is only of one sign, or polarity (conventionally called 'positive'). A massive body generates a gravitational potential, and its gradient is the gravitational field (Fig. 2.3C). The variations in the gravitational field are the familiar tidal forces. While the gravitational potential is a single quantity, the gravitational field has three components in the three directions, and its variations (tidal field) have many more components, like 'the variation of the y-component of the field in the z-direction in space', etc. That makes $3 \times 3 = 9$ components for the tidal field. Also, the x, y, z components can depend on time. Then there are $9 + 3 = 12$ components.

As an aside, these components of the tidal force are the components of what is called the 'curvature' of space-time in the General Theory of Relativity. Since, the GTR deals with time as one of the dimensions, there are 4 additional components, and 16 components in total for the curvature. Thus, the curvatures of space and time in the GTR are the familiar tidal fields, generalized to include 'time' as an additional dimension.

Now we focus on the gravitational potential, which is the important quantity for the modification of time (Fig. 2.4). For a small body of mass M, the potential is $V_g = -GM/r$ at a distance r away.

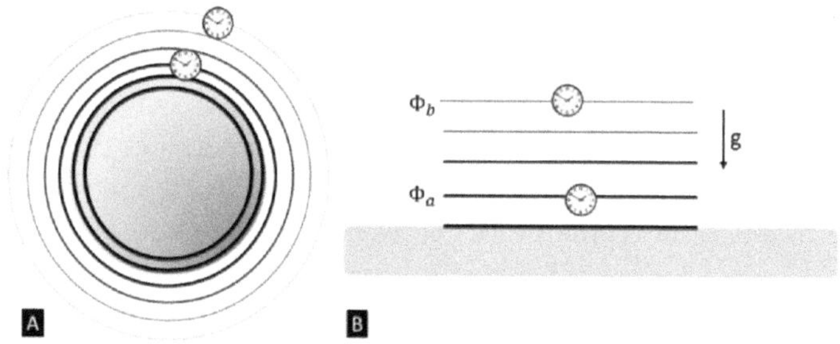

Figure 2.4 (A) For a spherical mass, both the gravitational field and the potential vary with distance. However, the rate of clocks depends on the value of the potential and not on the strength of the attractive force. (B) Near a large mass slab, the gravitational force is nearly a constant and does not vary with distance. But, the potential decreases more or less linearly as the distance increases. Two clocks in such constant gravitational field also have a relative time dilation because the potentials ϕ_a and ϕ_b are different.

The negative sign is chosen by convention, to agree with the fact that the force derived from this potential is attractive. (An attractive force is indicated by a negative sign, and the force is defined as the negative gradient of the potential.) Then the gravitational field is $g = -dV_g/dr = -GM/r^2$ and the pulling force on another body of mass m is $F_g = -GMm/r^2$.

The gravitational energy of a body of mass m in the gravitational potential of another body of mass M, separated by distance r, is $E_g = -GMm/r$. It is this energy that we convert to the electrical energy, etc., by allowing the body of mass m to 'fall' by some distance, thereby decreasing (making more negative) its gravitational energy, which can then be converted to the motional energy of something else, like a turbine.

What are the gravitational potential and the gravitational field inside a closed spherical vessel of mass M? The answer is very important for the central topic of this book. As one would have guessed, from our discussion on the electric potential and field in the case of a closed metal plate, the gravitational field inside a closed massive shell is expected to be zero. But gravity is not absent! When the field is zero, it just means that the gradient of the gravitational potential is zero. But the potential itself in this case is not zero; it is

Figure 2.5 Inside a very large volume of matter and energy we will not feel any gravitational force because we are in an 'equipotential space'. Therefore, if we do not have the tools to see the enormous amount of matter, we could be misled to think that our external space is mostly empty, 'without much gravity'. Note that on the Earth we are within a vast distribution of matter, extending to billions of light years, whose gravity we cannot 'feel' as a force. But the matter is observed and hence its enormous gravitational potential exists.

$-GM/R$ where R is the radius of the closed shell. If we are inside, we will not feel any force; we will be floating freely (Fig. 2.5). However, gravity is very much present and the gravitational energy of any mass m inside the shell is $E_g = -GMm/R$. This means that if there is any physical process that is modified by the gravitational potential energy, such modification would happen inside the shell, which can be detected when the physical system can be compared with another which was outside the shell, and far away from it. For example, everybody now knows that the rate of a clock is modified by a gravitational potential. *A clock inside the shell will have a different rate compared to another in free space, though both will not feel any force of gravity.* If these clocks are synchronized first and one is put inside a shell of large mass for a period of time (while the other is far away in free space), and then compared at mutual proximity, the clock that was inside the massive shell would show lesser time (aged less). That is, of course, the time dilation due to the gravitational potential.

Since it is better to strike the iron before it cools, I now state a subtle and profound fact.

> **Since the Universe has lots of matter that is distributed more or less uniformly and isotropically, each of us is in a situation similar to living inside a gigantic thick shell of matter, with no substantial gradient in the enormous gravitational potential of this matter. We do not feel any pulling force of this enormous gravity only because the variations of the potential are negligible, and not because there is no gravity! *There is no detectable gravitational field, but the gravitational potential is huge.* So, every physical process that is affected by a gravitational potential is getting affected significantly by this cosmic potential.**

We do not see this readily, because we are always comparing the change in one physical system (say, a clock) with that of another that is also inside this gigantic shell, in the same potential. Since all measurements are a comparison between a system and another reference system, we do not see any change when both are in fact changing similarly! We do not realize the cosmic gravitational effect on our clocks, and on our life and motion because it is not readily observable, unless of course we arrange to observe it. We will soon discuss how to observe the gravity of the Universe, which we fail to see because we are sensitive to only the forces, and not to the potentials. But our clocks and time have been significantly affected and modified by the gravity of the Universe. So have been many other physical phenomena, like the propagation of light. *Since we were not aware of the cosmic matter and its gravity, we misinterpreted our observations in terms of other fictitious constructs.* Not that we did not have empirical data – in fact, we had plenty of it in the last century because of the accelerated technological developments. But, the overarching fundamental theories were constructed before the facts about the matter distribution in the Universe were known. The cart was so ahead of the horses that every occasional guess of association was ignored, or even ridiculed. It is obvious that the physics of space, time and motion in empty space we built up over centuries is grossly incorrect and that we have to rebuild our physics afresh.

2.3.1 More on the Puzzle of Inertia

To realize the foundational depth of the Universality of Free Fall, consider motion under an electric force, which obeys a law similar to that of gravity – Coulomb's law of the electric force between a charge Q and another q,

$$\vec{F}_e = \hat{r}k\frac{Qq}{r^2} = \vec{f}q \tag{2.8}$$

If Q is a positive charge and q negative, the force is attractive. We can now determine the rate at which q accelerates and fall towards Q, using Newton's law, $\vec{a} = \vec{F}_g/m_i$. I have indicated explicitly that the 'mass' in the law of motion is the inertial mass (or inertia, for short), m_i. We get

$$\vec{a} = \vec{F}_e/m_i = \vec{f}\frac{q}{m_i} \tag{2.9}$$

This depends on the *ratio* of the electric charge to mass and varies from body to body. There is no universality.

In contrast, when we consider the gravitational force, the law of gravity involves the gravitational charges – the masses here are the gravitational masses M_g and m_g.

$$\vec{F}_g = -\hat{r}G\frac{M_g m_g}{r^2} = \vec{g}m_g \tag{2.10}$$

The law of motion remains the same, with the inertial mass, $\vec{a} = \vec{F}_g/m_i$. Then the gravitational acceleration is

$$\vec{a} = \vec{g}\frac{m_g}{m_i} \tag{2.11}$$

This depends on the ratio of the gravitational mass and the inertial mass. However, if the property of inertia is the same physical quantity as the gravitational charge, we get the universality of the gravitational acceleration

$$\vec{a} = \vec{g}\frac{m_g}{m_i} = \vec{g} \tag{2.12}$$

In fact, for the Universality of Free Fall, it will suffice if the ratio of the two quantities, m_g/m_i, is the same for all bodies, independent of the material properties.

The fundamental puzzle of why the inertia is universally related to the gravitational mass was never solved. It was finally sidelined by Einstein by assuming or postulating that the two quantities 'mass' and 'inertia', which could have been fundamentally different, are in fact the same. This is the Weak Equivalence Principle, which asserts the equivalence of the gravitational mass and the inertial mass. Einstein was a master of transforming a riddle into a worthy principle and then using it to derive new physics. Generalizing from the consequence of this, Einstein stated that a gravitational field $\vec{g}$ and an acceleration $\vec{a} = -\vec{g}$ generated by any means cannot be distinguished 'locally' with experiments within the local region. When stated in this general form, it is called the Einstein Equivalence Principle and it is the fundamental hypothesis and basis on which his theory of gravity, the General Theory of Relativity, is built.

Einstein wrote in 1914 (Physical foundations of a theory of gravitation) (Einstein, 1914),

> It is one of the most remarkable empirical facts of physics that these two masses, the inertial and the gravitational, agree exactly with each other as regards their magnitude...
>
> In a gravitational field all bodies fall with the same acceleration. This suggests the view that, with regard to its influence on mechanical and other physical processes, a gravitational field may be replaced by a state of acceleration of the reference body (coordinate system). This conception does not follow with necessity from the experiments mentioned, but it is of great heuristic interest all the same. For, since the course of physical processes relative to an accelerated reference system can be determined theoretically, this *equivalence hypothesis* permits us to predict the influence of a gravitational field on physical processes of every kind.

The basis of the founding hypothesis of the GTR, the equality of the inertial mass and the gravitational mass, is perhaps the deepest of puzzles in fundamental physics. By the time we progress to about half of this book, we will see the amazing solution to this fundamental puzzle. A solution was always in front of our eyes, and even speculated occasionally, but never taken seriously enough. When E. Mach argued that the reason for an inertial force like the centrifugal force might be some sort of interaction with the matter distributed in all space (Universe), it was a significant point to follow up (Mach, 1919). Einstein himself considered the possibility that the inertia of a body originates in the gravitational interaction with the rest of the matter in the Universe, and called it Mach's principle (Einstein, 1956). But he discarded the idea in favour of another conflicting principle called the general coordinate invariance to complete the GTR, presumably because there was no compelling reason, since nothing much was known about the matter in the Universe when the GTR was formulated. Later, some half-attempts were done, notably by D. Sciama, but it was not followed up to a firm conclusion (Sciama, 1953). Finally, the definite solution emerged from my re-examination of the entire scenario in the context of the factual cosmological observations and the reliable new knowledge about the amount and extent of the matter in the Universe. *It is indeed the gravitational interaction with all the matter in the Universe that defines the inertia of every piece of matter*. However, the gravity of the cosmic matter affects every aspect of fundamental physics, not just inertia. This paradigm-changing discovery and its consequences to relativity and dynamics are the contents of my theory of Cosmic Relativity (Unnikrishnan, 2004, 2009, 2020a). We will see that it is the relation between cosmic gravity and inertia that determines the motional time dilation in relativity. Our detailed examination of the experimental evidence on the behaviour of clocks will leave no room for doubt and uncertainty about the verity of the solution.

The term 'force' is just a proxy for the spatial variation of the energy of an interaction. The primary physical quantity is the energy of interaction between two objects. If this energy changes as the distance between the objects is changed, then it is called a force, and this is what living beings are sensitive to because their senses detect only differences and gradients. As we discussed already, the energy of interaction itself is not detected directly. This is crucially important for our discussion because we remain insensitive, not being able to experience by our sense organs, to the gravity of the uniform matter distribution enveloping us. The fact that there is indeed such an envelope of matter, our vast matter-filled Universe, was not known to us because of this sense-blindness, in spite of the fact that the actual interaction energy is a billion times larger than the energy of interaction between us and the Earth. Thus, the realization of the matter-filled Universe had to wait till we saw with our own eyes the galaxies distributed all around, and far, through the big telescopes of the 20th century.

However, there are physical entities that are sensitive to the interaction energy directly. A concrete example is a clock, marking time. Once the frequency of an oscillator is fixed, its 'phase' or the amount by which it has advanced in its periodic motion, is identical to time. The phase (of a wavefunction) is also the primary quantity that determines dynamics in the quantum theory. In quantum mechanics, energy is related to a frequency ($E = h\nu$) and the energy multiplied by the duration in dynamics is the same as the phase. Hence, any interaction energy, and in particular the energy of the gravitational interaction, directly translates as a modification of time.

2.4 Gravity's Charge

Gravity is always attractive. Therefore, there was no compelling need to characterize the nature of the source of gravity in any detail. All bodies generate gravity proportional to their mass, with the strength of gravity reducing as the square of the distance from the body; that was sufficient understanding of gravity, unlike electrodynamics where one finds both attraction and repulsion in forces. So, electrodynamics had to deal with two kinds of sources and they were termed positive and negative charges. Later, when magnetism was studied in detail, there was also positive and negative currents of these charges. Charge is just another name for the source of a force field. Mass is the source and hence mass is the 'charge of gravity'. In fact, the close resemblance of Coulomb's law of the electric force and Newton's law of gravitation is most striking when one learns elementary physics.

Relativity equates mass and energy through the relation $E = mc^2$, and hence, all forms of energy are also sources of gravity. *Since energy is equivalent to mass, energy is also inertia*. So, both mass and energy play the dual role as the source, or charge, of gravity and also as the inertia to motion. Therefore, the more one accelerates a body, thereby increasing its kinetic energy, the harder it becomes to accelerate it further, because of the increased inertia from the increased energy. Electric charge does not have such a double role. So, mass (and energy) have a special place in fundamental physics. Yet, as I already mentioned, present day physics does not understand why mass has this double role. All one could do was to *posit* that the two quantities of the gravitational charge and inertia are one and the same physical quantity we call 'mass'. That is the postulate of the (Weak) Equivalence Principle (WEP). Unravelling this fundamental puzzle was crucial in understanding dynamics, relativity, and indeed, the nature of time and its intimate relation with gravity.

2.5 Gravity as a Modification of Space and Time

Einstein's General Theory of Relativity (GTR) does not view gravity as a force field at all. The GTR starts with a clean and empty space with Euclidean geometry (flat) and time is added as another dimension of generalized coordinates, making a four-dimensional space-time, in which particles move along straight lines. (Adding time makes the geometry 'non-Euclidean', but that is a mathematical aspect we will defer for later mention.) The presence of matter alters the geometry of space-time and then a general 4D Riemannian curved geometry offers the correct description. Particles continue to move along the geodesics, or straightest trajectories, in the modified geometry. The geometrical features like the components of the curvature are related to the matter-energy density and momentum through Einstein's equation of the GTR (Taylor and Wheeler, 1992; Price, 1982; Einstein, 1920, 1956).

We already saw that the curvature in the GTR is factually the tidal forces, mostly. However, since gravitational potentials affect clocks, the definition of tidal field gets enlarged when time is added to space as an additional dimension. The 12-component tidal force in space (including their time dependence) becomes a 16-component tidal force in full space-time. This is what is called the 'curvature of space-time' in Einstein's theory. Why is it done this way? There is a valid physical reason. The potential itself is not directly detectable. The gravitational field is sensible, but apparently vanishes if we are in a spacecraft, or in a freely falling lift, or inside a spherical massive shell. The only measure of gravity that is present and sensible irrespective of the situation of motion or coordinates, etc., is the tidal distortions of spatial and temporal intervals. A tidal force does not vanish in an orbiting space station. That is the reason for the construction of the GTR in terms of the tidal forces, or the curvature (of space-time).

Let us recapitulate. A massive body in space, like the Earth, is a source of gravity. If there are other small bodies near such a source, their usually straight trajectories of free motion will be affected. First of all, the acceleration of these bodies does not depend on their mass or composition; there is universality in free fall. Since all bodies follow the same trajectory, from light to satellites, the properties of the bodies disappear from the description of dynamics, and we can say that it is the space itself that is affected, or 'curved', by the source; the bodies continue to move in *straightest lines in the curved space*.

As we saw already, due to the tidal forces, two free-fall trajectories separated by a short distance progressively come closer or farther, depending on the direction of the separation, as the bodies fall towards the Earth (Fig. 2.3B). This is exactly the character of the trajectories in a curved space, like two trajectories that start off as parallels on the surface of a sphere. The 'curvature' in the GTR is more complicated to describe than the curvature of a ball because we are dealing with 3 spatial dimensions and 1 extra dimension for time. Instead of saying that material bodies are universally affected by the gravity of a source body, the GTR describes this in two steps: the source affects space and time, and then the modified space-time is the 'reason' for the modified and curved trajectories of moving bodies. While achieving mathematical consistency, it pays the price of positing the unobservable metaphysical entities, space and time, as an intermediary for the action of gravitation.

In the space-time picture, gravity is very different from other interactions. The physical reality of the insensible space and time is postulated first and then gravity is equated to the curved geometry of that space-time, induced by matter. *There is no gravitational field or potential in the GTR.* As an immediate consequence, it becomes inconsistent to talk about the analogue of the photons of electromagnetism in gravity, even if we are talking about gravitational waves. There cannot be any particle corresponding to the gravitational waves because, in the GTR, the waves are just the perturbations of the curvature of space-time and not particles that are emitted by a source. In fact, most physicists talk about gravitational waves with the decorative description, 'ripples in the fabric of space-time'. This assumes the physical reality of an elastic four-dimensional medium, as a physical entity, through which the waves of distortions of the 'fabric' propagate. Clearly, one cannot get particles of any sort when the rules of quantum mechanics are applied to such distortions in space and time. In a hypothetical quantum theory of gravity based on the GTR there are no gravitons, if one is rigorous and truthful. Quantized space-time cannot be equated to particles of energy resembling photons. But, we need not dwell on such obvious problems because the quantum features of gravity are not relevant for our discussion of the relation between physical time and gravity.

Even the mathematical details of the GTR are not important for our study of time. The only result that is crucial for us is the change in the rate of clocks near a massive body, derived by Einstein well before he had a complete theory of gravity. We have already seen this derivation and the simple formula for the phenomenon. The gravitational time dilation of a clock, when the gravitational potential $\phi \ll c^2$, is given by the relation, $\Delta T/T = \phi/c^2$.

Einstein's equation of the GTR equates the generalized tidal features, described as the curvature of space and time, to the density and pressure of matter, with a proportionality constant that involves both the gravitational constant and the speed of light. For completeness, I display this equation:

$$R_{ik} - \frac{1}{2}g_{ik}R = \frac{8\pi G}{c^4}T_{ik} \tag{2.13}$$

R_{ik} and R represent the features of the curvature and g_{ik} is the 'metric', which again encodes the geometric features of space and time. The T_{ik} on the right side is the energy-momentum tensor of matter, with density, pressure, etc., as its components. The single equation represents many equations, due to the different components of the quantities involved. In effect, the curvature R_{ik} takes the role of the tidal forces and the metric g_{ik}, that of the gravitational potentials. The curvature of space-time involves the second derivatives of the metric. Therefore, Einstein's equation may be seen as a grand generalization of Poisson's equation of Newtonian gravity, $\nabla^2\phi = 4\pi G\rho$, where ρ is the matter density.

As practiced in present day physics, Einstein's equation simplifies to $R_{ik} = 0$ when one is sufficiently far away from localized matter like the Earth or the solar system, where there is no matter or energy (then T_{ik} is zero). This is called the vacuum Einstein equation (Price, 1982). *One can see very plainly that the fundamental equation for gravity and dynamics does not include the cosmic matter and its gravity, which are always present and cannot be removed.* There is nowhere in the Universe one could go and be sufficiently far away from matter and its gravity! Thus, the vacuum Einstein equation is an explicit reminder of the forgotten dominant cosmic matter and its gravity, in the current theory of gravity and relativity.

2.6 How Can Gravity Act on Void?

The current theory of gravity puts too many metaphysical loads on gravity as a physical phenomenon. Gravity is sourced by matter, with its mass as the charge. In the Newtonian theory of gravity, matter affects matter through a gravitational field, exactly similar to how an electric charge affects another through its electric field. The General Theory of Relativity, on the other hand, insists that gravity is matter affecting and modifying space and time. But, even after granting the unverifiable physicality of the unobservable space and time, one is confronted with the fact that they have no gravitational charge. Yet, the GTR dictates that space and time, without any gravitational charge of their own, are affected by the gravitational charge of matter. The gravitational constant denotes the ability of matter to modify space and time.

But there is also dynamics without gravity – matter moves and responds to electromagnetic fields without any reference to gravity. Then, dynamics has an existence separate from gravity. This implies that *the role of mass as inertia is independent of the gravitational interaction in the current scenario.* Electromagnetic forces do not modify space or time and yet, trajectories of motion deviate from straight lines in the presence of such forces. Whether space and time are modified by the presence of matter is a metaphysical question because this cannot be determined empirically due to the insensible nature of space and time. All we have observational access to is the response of matter to the presence of other matter. We can create a description in terms of the forces and responses, or in terms of the geodesics (trajectories) in a modified geometry. One description cannot be dismissed in favour of the other by a logical or empirical process. In fact, the empirical determination favours the physical picture of the interaction energy and force, because space-time remains hidden as a metaphysical cover within the empirical method.

An example might help to gain clarity. We know that changes in temperature make things expand or contract. Typically heated objects expand. But this is not universal. There are even objects that contract when heated! And there are engineered ones that do not expand much in some temperature range. So, it all depends on the nature of the material. However, suppose that all objects had the same thermal expansion, for some unknown reason. Then, we cannot sense or measure the thermal expansion of one object against another, used as a scale, and some other method of measurement would be required. In such a situation we could even postulate that the temperature expands space itself, and not materials! Materials just expand with space in which they are embedded. This example is convoluted, but it is fully representable in a mathematical theory.

The relation between matter and gravity is replaced in the GTR by the relation between matter and space, as well as time, in a manner similar to the example of universal thermal expansion. Since dynamics (motion) is described in terms of the distance of some matter from a reference grid made of matter, as a correlation with a clock (time), the entire situation can focus on the distance, rather than on matter; then one can even say that space (distances) and time are affected, and not the material body or the material clock, by gravity. However, the conceptual dangers of forgetting matter, which is the only entity that is physical and perceptible to observations, should be clear.

A person approaching physics as a science that concerns the physical world of matter does not fail to see the close similarity between electromagnetism (EM) and gravity, and also their differences. Both interactions are long range, with an 'inverse square law' for the forces. EM is sourced by electric charges and their currents. Magnetism is the interaction between currents. Similarly, gravity is sourced by massive matter and its currents. In fact, the GTR revealed that moving masses create additional gravitational fields very similar to the magnetic fields; only, they are too weak to be seen in common terrestrial situations. The new kind of gravitational field from the current of matter is called a 'gravitomagnetic field' (Ciufolini and Wheeler, 1995). However, instead of bringing the two theories close, Einstein's theory created a break between the two, blocking an integrated view. In the present view, EM acts in a space and time distorted by gravity, whereas EM does not affect space or time directly. However, the physical effects of the two are very similar – potentials that change the quantum dynamics of particles, forces that pull (or push) particles, tidal forces that distort matter, etc. A rotating charge has a magnetic moment that responds to a magnetic field, and a rotating mass has a 'spin' that responds similarly in a gravitomagnetic field created by other rotating masses or currents of masses.

One might vehemently object to bringing EM and gravity together in this manner, reducing the mysterious superiority that gravity enjoys in the present scheme of things. After all, gravity can affect clocks, but EM apparently can't. So, there must be something very special about gravity.

Therefore, if we can find out why gravity alone affects the rate of a clock whereas EM doesn't, perhaps we will gain better insight about gravity. The answer turns out to be simple, if we do not lose sight of what time really is. *Time is identical to an evolving physical process, and every physical process involves the mass (or energy) of matter. Therefore, every physical process and time itself are affected by gravity.* In contrast, every physical process does not involve a charged particle. In fact, when a charged particle is involved, the phenomenon is indeed affected by EM fields. Clocks, including atomic clocks, are easily perturbed by strong electric and magnetic fields. But, there is no universal influence of EM fields in physical processes. We will see this crucial difference in some detail, and in quantitative terms, later.

To summarize, EM fields are generated by electric charges and their currents. Then other charges and currents 'couple' to these fields and respond by altering their motion, shape, dynamics, etc. In contrast, masses and their currents, according to the GTR, do not generate any fields (strange, isn't it?). Instead, their presence alters the character of 'space and time'. The presence of matter creates a 'curvature' in space and time, and then the general dynamics of matter changes because the motion is now in the altered space and time. That is the theory of gravity today. However, far away from isolated matter like the Earth, Sun, or some star, or a black hole, space and time are 'flat', with no (or negligible) curvature. Then, there are no gravitational effects. Also, since one floats freely in a spacecraft or in a falling box in a gravitational field, one can think that there is no gravity in the neighbourhood. The GTR decrees that sufficiently far away from masses like the Earth, or in the small confines of a spacecraft, there is no gravity, and that the local space-time is flat – then the Special Theory of Relativity applies.

However, at this point the GTR fails to see the forest for the isolated trees! The GTR does not include the distributed matter in the Universe in its structure. It works with what is called the vacuum Einstein equation in such cases, whereas factually, there is no place one can go far enough from the Universe to reach the empty space of Special Relativity! It is this gross inconsistency that made the GTR miss several fundamental results, like the origin of inertia and inertial forces, the physical reason for the motional time dilation, the Galilean nature of the propagation of light, etc., and rendered the impressive theory of the GTR *incomplete*. Once the dominant gravitational presence of the cosmic matter is acknowledged and incorporated, as they are indeed in Cosmic Relativity, a very different picture of the physical world emerges, which is also the one that is overwhelmingly supported by the empirical results. We will see some of these deep results in the subsequent chapters.

2.7 Summary

In this chapter, we gained familiarity with the universal interaction of gravitation. We examined the different facets of the interaction, and recognized its physical structure in terms of the gravitational potential (interaction energy), the gravitational field (the pulling force), and the tidal force fields. Then we examined how the phenomenon of gravitational time dilation is deduced from Einstein's equivalence principle, and derived the formula for the dilation. The double role of mass as the charge, or the source of gravity, and also as the inertia for motion, was stressed and elaborated. We examined some aspects of describing gravity in terms of the modifications of space and time themselves, instead of as an interaction between material particles that affects the spatial and temporal intervals between material bodies.

I sketched the contents of the Weak Equivalence Principle, the Universality of Free Fall, and the Einstein Equivalence Principle. Their role and importance in the formulation of the GTR were discussed. Einstein's equation of the GTR was introduced, with some critical comments on the outstanding unexplained issues like the origin of inertia.

I mentioned the important fact that there is a vast amount of matter in the Universe, which is the source of a large gravitational potential. In other words, the interaction energy of every piece of matter with all the matter and energy in the Universe is enormous, compared to the interaction energy with local masses like the Earth or the Sun. However, since this fact was not known when our fundamental theories were constructed and completed in the early 1900s, our theories do not include this dominant gravitational interaction with the rest of the Universe, which nevertheless affects clocks in motion significantly. It is by studying this new aspect, which is beyond our current theoretical understanding, that we will realize very deep and new aspects about time and its modifications.

Recognizing the importance of the factual cosmic matter and its gravity to several aspects of fundamental physics is the basis of my theory of Cosmic Relativity, which is the complete cosmic gravitational paradigm for relativity and dynamics. Cosmic Relativity provides the correct and accurate description of the modification of time in moving clocks, as we will see in detail.

Since the Special Theory of Relativity and its well-known time dilation of moving clocks preceded the General Theory of Relativity by a decade, the modification of time during relative motion is widely discussed. This is a vital physical phenomenon that requires attentive discussion because of its subtle nature, and alive as the source of numerous paradoxes, debates, and confusion. We take up the study of the relation between motion and time in the next chapter.

Chapter 3

Time and Motion

The most well-known modification of physical time is the time dilation of moving clocks in relativity. This is familiar to everyone, being a topic of discussion even in popular literature and films. Slowing down the passage of time significantly is the only hope for even cosmic explorers in fiction to reach the distant worlds in our Universe, within a lifetime.

The topic of time dilation became an intense topic of discussion and debate within a few years of the publication of the Special Theory of Relativity (STR) by Einstein in 1905 (Einstein, 1905), in which he wrote,

> If one of two synchronous clocks at A is moved in a closed curve with constant velocity until it returns to A, the journey lasting t seconds, then by the clock which has remained at rest the travelled clock on its arrival at A will be $tv^2/2c^2$ second slow.

The phenomenon was mentioned by J. Larmor earlier in the context of the periodic motion of electrons. However, Einstein's explicit and clear prediction of time dilation of transported clocks was the first general statement on the motional modification of time as a physical phenomenon observable in any type of clock. It became

Gravity's Time
C. S. Unnikrishnan

ISBN 978-981-4877-72-5 (Hardcover), 978-1-003-25656-4 (eBook)
www.jennystanford.com

a topic of debate not because the phenomenon was radically strange, but because of the structure of the STR, in which motion was defined only in terms of *relative movement*; even the kinematical impression of motion was treated as physical motion in the STR.

The study of relativity is the study of physical effects during relative motion. Before Einstein, there were two important insights gained by physicists on relativity. One was Galileo's realization that the state of uniform motion in a straight line could not be distinguished from the state of rest, by any physical means or experiments. This is distilled from our everyday experience; we often face the situation of the apparition that we are moving, while in fact only an external reference background is moving, as with the slowly moving trains in a station. On the other hand, it is a common and familiar situation that we cannot sense our smooth motion, even when it is at a tremendous speed. We do not feel, and cannot prove, the fast motion of the Earth around the Sun, to the extent of not even believing it for centuries. Therefore, it is reasonable to state that only relative motion makes physical sense. Anybody in uniform motion can claim a state of rest, ascribing the whole motion to the rest of the world. *All laws of physics should conform to this 'Principle of Relativity' (PoR).* In other words, the laws of physics are the same in every frame that is in uniform motion. Uniform rectilinear motion is commonly called inertial motion. There is no acceleration in ideal inertial motion, $\vec{a} = d\vec{v}/dt = 0$.

Newton's law of motion $\vec{a} = \vec{F}/m$ certainly does obey the PoR. Since in inertial motion the velocity v is constant in magnitude and direction, the acceleration $\vec{a} = d\vec{v}/dt = 0$. Therefore, inertial motion cannot be sensed by any device that obeys Newton's law of motion, which is the statement of the Galilean PoR.

The second important insight about inertial motion was that such motion could not be detected even by any physical means that went beyond Newton's laws, like the propagation of light and the general electromagnetic phenomena. The 'ether' was the hypothetical medium that supported the propagation of light as waves. Progressively it became clear that the electromagnetic phenomena also obeyed the PoR, and that it was a general principle of nature. This realization was relatively more recent, stated only towards the end of the 19th century. It is in this context that physicists were forced to introduce the phenomenon of the velocity dependent modification of time, as well as of spatial distances. Reconciling new experimental evidence required the radical postulate that spatial lengths contracted and time intervals stretched, when in motion relative to the ether medium. Several researchers – G. F. Fitzgerald, H. A. Lorentz, H. Poincaré, and J. Larmor – contributed, before Lorentz and Poincaré formulated the strange phenomena in a mathematically comprehensive theory of relativity (Whittaker, 1954). Since time is an accumulated quantity, the accrued time dilations remain even after the end of motion, amenable to convenient measurements and comparison at rest. This was the first instance in physics that the passage of the physical time, assumed till then as beyond any human manipulation, could be modified by mere motion. Time became subjective, one stream for each observer and reference frame, with a multitude of real times that depended on the speed of motion of the clock. However, one serious fundamental problem remained; according to the PoR, the inertial motion relative to the ether could not be detected! Hence, there was no way of knowing or predicting the expected modifications, in spite of the ready formulae.

Einstein's Special Theory of Relativity had the same physical effects, length contraction and time dilation. However, *there was a crucial difference.* The relativity of Lorentz and Poincaré implemented the PoR in the hypothetical medium that supported the electromagnetic phenomena and the propagation of light. This meant that all relative motion was relative to the single all-pervading medium of the ether. Though invisible and insensible, the ether was acting as an *absolute reference frame,* akin to Newton's absolute space. Thus, there was at least a notional meaning to the state of absolute rest; a state of rest relative to the ether would be true and absolute rest. The propagation of light had a fixed speed relative to the ether and the relative velocity of light varied according to the velocity of the reference frame of the observer, similar to how the relative velocity of sound changes depending on the velocity of the observer relative to the medium of air. In contrast, Einstein implemented the PoR in empty space, devoid of the ether or any notional reference with an absolute nature. Thus, there was no notion of absolute rest either. *All inertial observers were equivalent and any inertial observer could claim the state of rest with full right.* For consistency, Einstein had to make the radical assumption of the invariance of the relative velocity of light for any observer in inertial motion.

We will examine the relevant details of Einstein's theory and its important conclusions about time in this chapter. In the process, we will be very attentive about the logical and the empirical aspects. We will also make relevant comments about the points of tension between the physics of relativity and our familiar physical intuition, even though we will not depend on 'common sense' as a criterion for the evaluation of the physical theory. After all, the physical world could be stranger than what our common sense could accept.

3.1 Relativity and the Propagation of Light

The development of the ideas on relativity is historically intertwined with the study of light. The discovery of the modification of time, with a quantitative measure that depends on the speed of light, was of course unexpected. Though the modern relativity theories came about initially as linked to the electromagnetic phenomena, the canvas was more general in the STR, pertaining to the relativity of all physical phenomena. Therefore, the prominent role of the 'speed of light' in the relativistic expressions and the entire mathematical structure of relativity remained as a puzzle. The STR had an upper limit for the velocity of any motion $v_{\max}$, including the propagation of signals and interactions. This turned out to be equal to the speed of light c. The speed of light is determined by the constant parameters that characterize the electromagnetic interaction, as $c^2 = 1/\epsilon_0\mu_0$, where ϵ_0 is the electric permittivity constant and μ_0 is the magnetic permeability constant. Why should the maximum possible velocity of motion and interactions in empty space be related to the parameters specific to the electromagnetic interaction?! It would have been physically more reasonable if $v_{\max}$ was determined by some fact that was more universal than electromagnetism. Notwithstanding this puzzle, it is a historical fact that the entire development of the theories of relativity was centred on the propagation of light.

From ancient days, the human notion of time is tightly linked to the leading light of the Sun. But, that is only a gross link between time and light, mostly in the sense of the division between day and night.

Our primary and only senses of the distant world are vision and hearing. The rest of our senses are 'short range'. Compared to sound, light indicates a superior power of intensity, speed, and grandeur to us. Light is our only link to the larger Universe. In comparison to sound, light seemed instantaneous. In a sense, light seemed beyond time – a thoroughly worshipable trait!

The idea that light had a finite speed of propagation occurred to many. Galileo even tried to measure it by placing an assistant far away. However, the methods and resources available then were too crude to see light's enormous 'finiteness'. The first measurement of the speed of light was finally done from the observations on the eclipses of the satellites of Jupiter, by O. Rømer in about 1676.

By the 19th century, with both optical and mechanical techniques advancing, the finite speed of light could be measured in the laboratory. These techniques also suggested that extremely small durations of time could be marked and measured by comparison with a 'light clock', in which light bounced between two fixed surfaces.

By mid-19th century, it was becoming increasingly clear that there was a close relation between the speed of light and the electromagnetic (EM) quantities. When Maxwell completed his mathematical theory of electromagnetism, from which the wave equation for the propagation of electromagnetic waves with their velocity of propagation equal to the speed of light emerged, there was no remaining doubt (Maxwell, 1954). Confirmation followed soon, with the results from H. Hertz's pioneering experiments.

3.2 Relativity and Special Relativity

The definite completion of the theory of electrodynamics and the identification of light as the propagating waves of EM fields were the high point in 19th century physics. It was assumed that the propagation needs a universal medium, called the ether. The Galilean principle of relativity, that the state of uniform motion cannot be detected and distinguished from a state of rest, was at the basis of dynamics. This is related to the undetectability of pure 'space'. The question naturally arose whether the hypothetical ether could be detected by measuring the relative velocity of light while the laboratory on the Earth was moving through the ether, just as one could detect the presence of the medium of air by measuring the velocity of sound relative to an observer moving and chasing the sound waves. However, the enormous velocity of light makes this task practically very difficult. The solution employing an optical interferometer was invented by A. A. Michelson in 1880 (Michelson, 1927). The general idea is simple to understand.

Consider a pulse of light moving in a medium (ether) along the x-axis at the speed c. It takes the duration L/c to cover the distance L between the source and detector. If the apparatus and the observer are moving together relative to the medium along the x-axis at the speed v, chasing the light pulse, the relative speed of light is reduced, to $c - v$. Then the time taken to reach the detector depends on the velocity of the observer as

$$\Delta t_{-} = L/(c - v) \simeq (L/c)(1 + v/c) = (L/c) + Lv/c^2 \tag{3.1}$$

The additional delay in the propagation (Lv/c^2) that depends on the speed of the observer was thought to be practically impossible to measure, due to the enormous speed of light. For example, the duration of propagation (L/c) over 100 m is about 0.3 μs. The *additional delay* due to the motion of the Earth in its orbit at 30 km/s is only 30 ps. If the light pulse is brought back to the source point by reflection, the duration taken in the return trip is $\Delta t_{+} = L/(c + v)$, because light that is traveling against the motion of the apparatus has the relative speed $c + v$. Thus, the total duration is

$$\Delta t_x = \frac{L}{c - v} + \frac{L}{c + v} = \frac{2L}{c(1 - v^2/c^2)} \tag{3.2}$$

Now, the dependence on the speed of the apparatus has become second order in the velocity of motion, v^2/c^2. Even with v of the order of the Earth's orbital velocity of 30 km/s, v^2/c^2 is a tiny correction in the duration of propagation, of a part in 10^8. Then, for $L = 100$ m and $v = 30$ km/s, the delay in the *two-way propagation* varies by only about a femtosecond, which is nearly impossible to

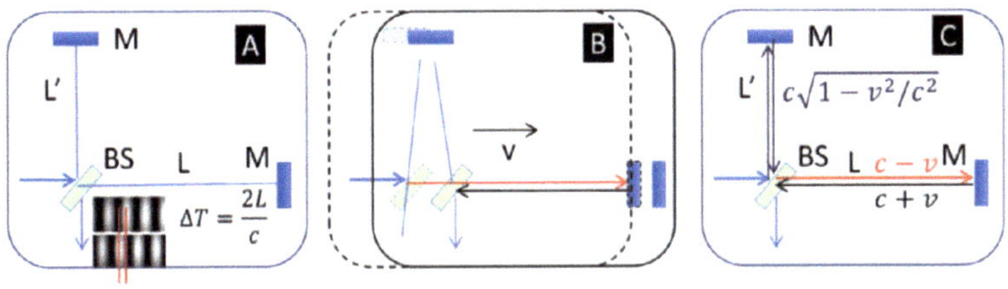

Figure 3.1 (A) The Michelson interferometer at rest (the original experiment had both the arms L and L' equal). Light travels from the beam splitter (BS) to the mirrors (M) and then reflects back to the BS. There, the two beams overlap and interfere, forming the bright and dark fringes. A change in the *difference* of the length of the arms, $L - L'$, shifts the bright and dark fringes. (B) The interferometer in motion at the velocity v, as observed from an external reference frame. Light travels different distances along and opposite the direction of motion. (C) The interferometer and the observer in motion together, and hence in relative rest. The external world and the 'ether' are relatively moving at the velocity $-v$. The relative speed of light one way is $c - v$, and $c + v$ during return. Along the perpendicular arm, the relative speed is $c\sqrt{1 - v^2/c^2}$.

measure directly even with today's technology. However, Michelson realized in 1880 that the wave property of light could be used to enormously improve the precision of such measurements.

Michelson devised an interferometer with two orthogonal arms (Fig. 3.1A). Michelson's interferometer had both the arms of equal length, $L' = L$. I have indicated the more general case with unequal arms, suitable for the general proofs of the points discussed. For the arm of the apparatus in the perpendicular direction (y-axis), the propagation delay over a length L is slightly different from what we calculated for the arm that is along the velocity v (he got good advice on the calculation from J. C. Maxwell). Light takes the duration $L/c\sqrt{1 - v^2/c^2}$ both ways along the perpendicular arm. The total two-way propagation duration along y-axis (Fig. 3.2) is

$$\Delta t_y = \frac{2L}{c\sqrt{1 - v^2/c^2}} \tag{3.3}$$

Though it is nearly impossible to measure such tiny corrections to the propagation duration at rest ($2L/c$), Michelson had the advantage of interferometry in which a 50:50 beam splitter ('half-silvered mirror') is used to split the light beam equally along the two

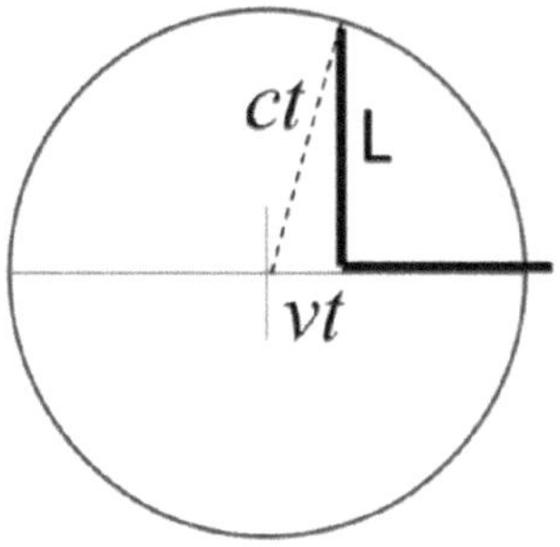

Figure 3.2 Light travels along the perpendicular arm of the Michelson interferometer a distance $L = \sqrt{(c^2t^2 - v^2t^2)} = ct\sqrt{(1 - v^2/c^2)}$ in duration t. Therefore, $t = L/c\sqrt{(1 - v^2/c^2)}$.

axes. After the reflection from mirrors, the beams propagate back to the beam splitter, where they are recombined. The superposed beams interfere and produce the interference fringes. A bright fringe will turn dark when the relative phase between the two beams changes by π, which is equal to half the wavelength, or about 300 nm. We note the very important fact that in such an experiment, there is no way to distinguish between an unequal change in the length of the two arms of the interferometer and a change in the duration of propagation of light. With refined measurement abilities of the master experimenter, Michelson's apparatus could sense a shift that was 100 times smaller. This corresponds to a difference in the propagation duration of $3 \times 10^{-9}/c \simeq 10^{-17}$ s, measurable owing to the large value of the speed of light! A seemingly inaccessible measurement of a tiny duration was made feasible by converting it to a measurement of an equivalent spatial interval. The innovative Michelson interferometer was sensitive enough, in principle, to easily detect the motion of the Earth through the ether medium, at its orbital speed around the Sun. In the late 19th century, the faster motion of the entire solar system at about 400 km/s was not known.

The difference in the duration of propagation along the two arms is

$$\Delta t_x - \Delta t_y = \frac{2L_x}{c(1 - v^2/c^2)} - \frac{2L_y}{c\sqrt{1 - v^2/c^2}} \simeq \frac{2L}{c}\frac{v^2}{2c^2} \tag{3.4}$$

We assume that the two arms are approximately equal, $L_x \simeq L_y$. Multiplying the difference in the duration of propagation by the (average) speed of light and dividing by half the wavelength gives the observable signal in terms of the shift of the interference fringes. The expected velocity-dependent signal in the experiment was

$$\delta n_f = \frac{2L}{\lambda}\frac{v^2}{2c^2} \tag{3.5}$$

With the distance L of several metres, the factor $2L/\lambda$ exceeds 10^7. With $v \simeq 30$ km/s, $n_f = 0.05$ or 1/20 of a fringe. If the apparatus is rotated, either by the experimenter or due to the rotation of the Earth, the arm that is along the motion of the Earth alternates and the fringes should shift.

From a refined experiment in 1887, Michelson and his collaborator E. W. Morley announced the failure to detect any shift of the fringes and the motion of the Earth through the ether, in spite of sufficient sensitivity of the experiment (Michelson, 1927; Michelson and Morley, 1887). *It was as if the ether was as undetectable as the empty space*; the Galilean principle of relativity seemed to be valid more generally, including all phenomena in mechanics and electrodynamics.

Both G. F. Fitzgerald and H. A. Lorentz independently thought of the same physical reason – that the length 'L' along the direction of motion *contracted* by the amount $\sqrt{(1 - v^2/c^2)}$. Then,

$$\Delta t_x = \frac{2L_x\sqrt{(1 - v^2/c^2)}}{c(1 - v^2/c^2)} = \frac{2L_x}{c\sqrt{(1 - v^2/c^2)}} \tag{3.6}$$

The length contraction makes both Δt_x and Δt_y equal and their difference, written in eq. 3.3, becomes zero! This could explain the null result of the Michelson–Morley (M-M) experiment. In other words, though the propagation of light is indeed Galilean, with relative velocity $c \pm v$, the physical phenomenon of length contraction makes the distance to propagate slightly smaller, hiding the Galilean nature of propagation, and preserving the PoR. *The M-M experiment showed that the PoR was valid in spite of the Galilean propagation of light, because of the phenomenon of length contraction.*

The null result of the M-M experiment implied the invariance of the Maxwell's equations of electrodynamics in uniformly moving frames, similar to the invariance or equivalence of Newton's laws in mechanics for all uniformly (inertially) moving observers. The

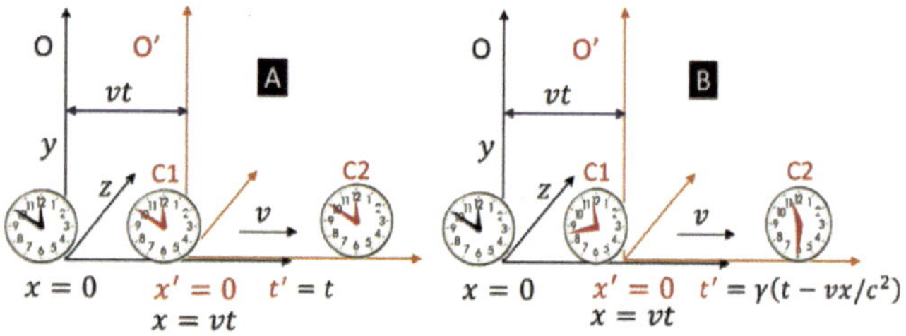

Figure 3.3 Schematic aspects of the (A) Galilean Transformations (GT) and the (B) Lorentz Transformations (LT). In the GT, the time coordinate is absolute. The clocks C1 and C2 in the moving frame O′ keep the same time as the clock in the stationary frame O. In the LT, the clocks at different locations in O′ have different time readings, relative to the clock in O. Also, all spatial distances along the direction of motion are contracted. In the STR, the LT are symmetric – either of O and O′ can be chosen as the stationary frame.

equations and their physical consequences were the same in all inertial frames. In other words, they could not be used to identify the velocity of inertial motion and distinguish the state of motion from the state of rest. However, the underlying mathematical relations between relatively moving frames were not the ones based on Galilean relativity, which were simple to understand. In the Galilean version, time was 'absolute' – the same for all observers. This is stated as $t' = t$, where t and t' are the time coordinates in the two frames O and O′. Since the frame O′ is moving at velocity v along the x-axis relative to O, the position x' in O′ is at the distance vt away at time t, from the coordinates x in O; $x = x' + vt$. So, the Galilean Transformation (GT) for the x-coordinate is $x' = x - vt$. The other coordinates do not change, $y' = y$ and $z' = z$ (Fig. 3.3A). The universal nature of time in Galilean relativity implies absolute simultaneity, by which two spatially separated events can be ordered in time, independent of the state of motion of the reference frame. It is as if the same master clock is available everywhere in the Universe. Then there is an invariant sense in deciding whether two distant events are simultaneous or not. However, it turned out that these coordinate relations could not reproduce the null result of the M-M experiment and the invariance of Maxwell's equations, even though what was seen was a general form of the Galilean principle of relativity.

Starting with the ad hoc hypothesis of length contraction (that all extensions contract in the direction of the motion by a certain fraction determined by the velocity through the ether), H. A. Lorentz eventually arrived at the 'Lorentz Transformations' (LT) of the spatial coordinates and time that achieved this invariance, while preserving the invisible ether and the Galilean notion of the variable relative speed of light (Lorentz, 1904). The Lorentz Transformations linking the space and time coordinates of a frame F' moving relative to a stationary (in the ether) frame F, with the velocity v along the x-axis are

$$x' = \gamma(x - vt) \tag{3.7}$$

$$y' = y; z' = z \tag{3.8}$$

$$t' = \gamma(t - vx/c^2) \tag{3.9}$$

Here, $\gamma = 1/(1 - v^2/c^2)^{1/2}$.

The difference from the familiar Galilean Transformations is this second-order (v^2/c^2) term, as well as the additional first-order term $(v/c)(x/c)$ in the transformation for time (Fig. 3.3B). When the velocity is small compared to the velocity of light, γ is tiny, and not relevant for most situations. Even then, there is a marked difference compared to the GT, by which $t' = t$. In particular, the LT do not reduce to the GT, contrary to common statements, when $v \ll c$, because of the x-dependent term vx/c^2. There is a general belief that the GT are the approximation to the LT, in the limit of the speed of light becoming infinite. This sounds reasonable, but it is incorrect! Formally, the LT do look like the GT when we put $c = \infty$. However, look at the Galilean formula for the Doppler shift of light when the source is moving towards the observer, $\nu' = \nu(1 + v/c)$. In the limit $c = \infty$, the Doppler shift vanishes! Therefore, it is clear that the GT are not the $c \rightarrow \infty$ limit of the LT.

It was the first-order term $-vx/c^2$ that was initially introduced by Lorentz, as a modification of the Galilean Transformations, to explain the fact that the motion of the Earth through the hypothetical ether could not be detected in any experiment that was sensitive to any first-order (v/c) physical effect. Stellar aberration is one such physical effect. The Maxwell equations of electrodynamics remain invariant to first order in v/c, under inertial motion, if the term vx/c^2 is subtracted from the time marked by a clock at the position x. This defines a new 'local time', $t' = t - vx/c^2$; 'local' because it depends in the location x. Lorentz considered this as a mathematical transformation without any physical implication for the operation of clocks at different positions. According to Lorentz, it was an adjustment to be made to the clocks at different positions to ensure the invariance of Maxwell's equation in a frame moving at the velocity v. The second-order correction $\gamma = 1/\sqrt{1 - v^2/c^2}$ was introduced in an effort to explain the null result of the M-M experiment.

From the first expression for x' in the LT, the length of an object, $dx' = x'_2 - x'_1$, measured in the frame F' (by an observer in F') is transformed to $dx = x_2 - x_1 = \sqrt{(1 - v^2/c^2)}dx'$ as measured by an observer in the frame F, relative to which F' is moving. Since $\sqrt{1 - v^2/c^2}$ is less than 1, all lengths contract in motion.

The Lorentz Transformation for time is more subtle. If a clock is carried at $x' = 0$ in the moving frame F', it is at different locations x_1 and x_2 relative to another frame F at the two instants t_1 and t_2, due to the motion. From the expression for the LT for time, we get for a small duration $dt' = t'_2 - t'_1$, during which the clock moves by $dx = vdt$,

$$dt' = \gamma(dt - vdx/c^2) = \frac{1}{\sqrt{1 - v^2/c^2}}dt(1 - v^2/c^2) = dt\sqrt{1 - v^2/c^2} \tag{3.10}$$

The duration in the moving clock is shorter compared to the duration measured in a clock that is stationary relative to the frame F. *Time passes slower in the moving clock.* Moving clocks tick slower. This is the relativistic time dilation.

In Lorentz's theory, the ether is like a universal absolute reference medium that remains undetectable. Real motion is relative to the ether. The resemblance to Newton's insensible, nevertheless 'real', absolute space is evident. A measuring rod that moves relative to the ether contracts and the time measured by a moving clock progresses slower. Henri Poincaré provided the decisive mathematical completion to this theory (Poincaré, 1905) and also clarified the particular term of 'local time' in the Lorentz Transformation of time, as linked to the synchronization of clocks at different places. We note that the transformation of the spatial coordinates is modified by only the second-order term γ, compared to the Galilean Transformations, whereas the transformation of the time coordinate has the additional position dependent local time, $-xv/c^2$, which is linear in v/c. Lorentz related the physical reason for the motional modifications of the spatial and temporal intervals to electromagnetic phenomena in the ether. Since extended matter is atomic matter held together by electromagnetic forces, as understood before the 20th century, this seemed reasonable then.

That was when Einstein published his ideas of relativity in 1905, with the principle of relativity as one of the postulates and a very different interpretation of the Lorentz Transformations. The different interpretation comes from the second and characteristic postulate of his theory – that the relative velocity of light is an invariant constant, relative to any moving observer (the light hypothesis). Once this is assumed, the Lorentz Transformations follow as a consequence. Of course, the null result of the M-M experiment is then easily explained; light takes duration $\Delta t = L/c$ each way, along both arms, independent of the velocity of the reference frame. Therefore, the difference in the durations and the relative phase of light at the detector remain constant. Then the fringes do not shift when the apparatus is rotated. Note the sharp contrast of the explanation of the null result of the M-M experiment: the propagation of light is at an invariant relative velocity, a constant that is independent of the motion of the reference frame. Hence, the duration of propagation is independent of whether the interferometer is moving or not, and the fringes do not shift even if the interferometer is rotated, or if its velocity changes.

All other consequences that Lorentz was struggling to understand as physically linked to the properties of the ether follow as mathematical consequences of Einstein's theory without the ether. But there was a special price to pay – without any universal reference, the theory became symmetrical and reciprocal between any two observers. *Any observer 'A' can claim the state of rest and that the other observer 'B' is moving, with the Lorentz modifications happening only to the clocks and rulers of the observer B. In turn, B can equally well claim that he is at rest, and A is moving with exactly the same modifications happening to A's clocks and rulers.* The fastest clock and the longest ruler are always in the frame at rest. But, the frame at rest is equally A and B, from their own frame. This is the point of departure from common sense and familiar world of space and time, in Einstein's theory of relativity. In other words, either A or B (or any other inertial frame) can claim the status of the special rest frame, which only the ether frame could claim earlier. This means that there is no meaning to the questions, 'which frame is really moving?', 'which clock and ruler are actually affected by motion?', etc. In the chosen rest frame, clocks and rulers are not affected. If this is seriously clashing with the intuitions and common sense built up over time, that is precisely the uneasiness that troubled many philosophers and even physicists.

Max Planck wrote about the PoR in Einstein's theory (Planck, 1915),

> The gist of this principle is: it is in no wise possible to detect the motion of a body relative to empty space; in fact, there is absolutely no physical sense in speaking of such a motion. If, therefore, two observers move with uniform but different velocities, then each of the two with exactly the same right may assert that with respect to empty space he is at rest, and there are no physical methods of measurement enabling us to decide in favour of the one or the other.

So, *in Einstein's theory, there is no physical reason for the modification of time and length.* It is totally symmetrical between two observers in relative motion. There is no way to claim or decide that a set of measuring rods and clocks in a particular frame suffers modifications due to motion. The frame chosen as the frame at rest is free from motional modifications. Strange as it may sound, what is demanded is only physical consistency within each reference frame, and not conformity to common sense or familiar logic. Since the observer A in one frame can only imagine and calculate what any other observer B measures in a different frame of reference, using the theory, but cannot actually measure using B's clocks and rulers (which are in another frame that is relatively moving), the mixing of the measurement results from the two frames does not happen. All A and B can do is to compare their measurements when they are at the same place – in infinitesimal proximity. Then of course, physical quantities should agree with the calculations in the theory, in both the frames. The discussions of whether and how this can be achieved are the sources of the widely debated paradoxes in Einstein's theory of relativity.

Nevertheless, Einstein emphasized the physical consequences of his theory in his paper in 1905. On the time dilation of clocks he wrote,

> ... it follows that the time marked by the clock (viewed in the stationary system) is slow by $1-\sqrt{1-v^2/c^2}$ seconds per second, or – neglecting magnitudes of fourth and higher order – by $v^2/2c^2$ ...
>
> If one of two synchronous clocks at A is moved in a closed curve with constant velocity until it returns to A, the journey lasting t seconds, then by the clock which has remained at rest the travelled clock on its arrival at A will be $tv^2/2c^2$ second slow.

This was a conceptually dramatic and radical consequence, though the amount of time dilation is minuscule. Even for the imaginable space-travel speeds of thousands of km/s, the time dilation is below a percent. But, nobody till then imagined that time could be malleable, and could be manipulated willfully by motion.

We see that *there were two mature theories of relativity in the early years of the 20th century; one was the Lorentz–Poincaré Theory of Relativity (LPTR) based on the absolute reference frame of the ether, and the other was Einstein's Special Theory of Relativity (STR) in empty space.* Both assumed the Principe of Relativity and both had the Lorentz Transformations as their mathematical structure. The STR had a second postulate, of the invariance of the relative speed of light in any inertial frame, which made the theory symmetric and equivalent between any two inertial frames. The LPTR did not share this feature. In the LPTR, the speed of light c was its speed in the ether frame, and in all other moving frames the relative speed depended on the velocity of the observer frame. The speed c was the limit of motion in the ether frame, but the speed relative to an observer could exceed c, similar to the fact that the relative speed of sound could exceed its value in the medium.

Interestingly, both theories agreed with all known experimental results. One could not favour one theory over the other based on any empirical result! We will not dwell on the multiple reasons for the universal eventual acceptance of Einstein's theory as the standard theory of relativity. However, it is important to state very clearly that to date there is not a single experimental result that favours the STR over the LPTR. We will now focus on the motional modification of clocks and time, and on the various aspects of its measurements and verification.

Since my firm assertion that the M-M experiment or its variations cannot decide between the STR and the LPTR (or a theory with an absolute frame) will be a surprise even to many expert physicists, I will present a simple proof for the claim. Of course, the fact that Lorentz could explain the null result, in a theory with the ether medium for the propagation of light, in which the relative velocity of light was explicitly Galilean ($c \pm v$), should be a sufficient proof. However, a focused proof for the general case will be definitive.

Assume explicitly that the relative speed of light is indeed Galilean, $c_r = c \pm v$. We are in the frame of the interferometer. We let the lengths of the two arms to be different, as in a general case. The total duration for the two-way propagation is already seen to be

$$\Delta T = \Delta t_x - \Delta t_y = \frac{2L_x}{c(1 - v^2/c^2)} - \frac{2L_y}{c\sqrt{1 - v^2/c^2}} \tag{3.11}$$

The two arms were equal, $L_x = L_y$ in the M-M experiment. The situation where $L_y \ll L_x$ is that of the Kennedy–Thorndike experiment (Kennedy and Thorndike, 1932), and the case $L_y = 0$ resembles that of an optical cavity (when the beam splitter is replaced by a mirror). At rest, light would have taken the duration $T = 2L_x/c$. But, in motion, light takes more time, $T/(1 - v^2/c^2)$. Since the length contraction and time dilation are real physical effects in the LPTR, due to the motion relative to the ether, each effect contributes the factor $\sqrt{1 - v^2/c^2}$ in the first term in the equation 3.11: length is contracted as $L_x\sqrt{1 - v^2/c^2}$. Then light has to propagate over a shorter distance, which changes the duration to propagate the distance L_x to $T' = T/\sqrt{(1 - v^2/c^2)}$. Additionally, since time progresses slower by the factor $\sqrt{1 - v^2/c^2}$, the final duration is time dilated, $T'\sqrt{(1-v^2/c^2)} = T = 2L_x/c$, which is the same as the duration at rest.

For the perpendicular arm, there is no length contraction, but there is time dilation. The time dilation factor $\sqrt{1-v^2/c^2}$ cancels the excess duration of propagation in Δt_y. We see that *the propagation duration in each arm is independently equal to its value at absolute rest, due to the physical length contraction and time dilation, in spite of explicitly assuming the Galilean propagation of light with the relative velocity* $c \pm v$. Therefore, the M-M experiment and its variants cannot demarcate between the STR and the LPTR, contrary to the belief held by most physicists. Clearly, the often-made statement that the M-M experiment is the experimental evidence for the constancy and isotropy of the relative velocity of light is grossly incorrect. As we have seen, the proof is elementary and there is no room for any ambiguity.

In spite of the transparent nature of this simple proof, most people believe that the M-M experiment is the experimental proof of the constancy of the relative velocity of light. This might be because most do not realize that such an experiment involves both light and a material apparatus, and that the null result could be a result of some modification happening to the apparatus in motion, rather than due to an invariance property of light! This is discussed in detail in the next chapter that deals with the principle of relativity.

The situation is somewhat like the inability to distinguish and demarcate between two colours when one is severely colour blind! Both hues of explanations equally fit the observed results. The only way out is to devise a method that does not suffer from the fundamental inability of the two-way experiments to demarcate; there is only a second-order signal (v^2/c^2) in such experiments in which light propagates back and forth, along and against the velocity of the frame. One immediate thought is to devise a 'one-way experiment', with a first-order (v/c) signal, but there are subtle and fundamental issues to be solved, as we will see in detail later.

In fact, the physicists of the 'relativity generation' (say, 1910–1950) were well aware that the postulate of the invariant relative speed of light was merely a consistent postulate and not an empirically proven fact. It is instructive to note that Michelson himself did not even mention the M-M experiment in the chapter dealing with the 'velocity of light', in his monograph, *Studies in Optics* (Michelson, 1927). It is discussed in the chapter on relativity, where his analysis of the null result did in fact assume a stationary ether. As another example, I quote from the notes by the physicist-astronomer Charles Nordmann, written on the occasion of Einstein's visit to Paris in 1922 (Nordmann, 1922),

> However, . . . there is still something infinitely troubling in the Einsteinian system. This system is admirably coherent, but it rests on a particular conception of the propagation of light. How are we to imagine that the propagation of a ray of light could be identical for an observer who flies away from it, and for an observer who rushes forward to meet it? If this is possible, it is in any case inconceivable to our customary mentality, and no matter how hard we try, we cannot make the mechanism and nature of that propagation intelligible.
>
> It must be confessed that here lies a mystery which eludes us. The whole Einsteinian synthesis, as coherent as it is, rests on a mystery, exactly like the revealed religions.

I also note, with considerable emphasis, that both these theories (STR and LPTR) were formulated in a space devoid of matter and energy, and their gravity, because there was no awareness and knowledge about the enormous amount of matter in the Universe, in the early 20th century. Therefore, the important cosmic gravitational effects on clocks have been totally missing in both the LPTR and the STR. This will turn out to be the crucial element that decides the correct theory of relativity. As one can sense, the consideration of the factually large cosmic gravitational effect on clocks will prove that neither of these theories was adequate. When cosmic gravity is recognized as the main physical factor in the motional modifications of spatial and temporal intervals, we will possess the correct theory of relativity, conforming to reason, logic, empirical results, and even physical intuition, built over 400 years of renaissance physics. Since we will have no difficulty to accept the observed reality of the vast amount of matter in the Universe, we will not face any conceptual or logical hurdle in understanding the new gravitational paradigm for the theory of relativity, when I discuss the theory of 'Cosmic Relativity' in the next chapter.

3.3 Motion and Physical Common Sense

The Principle of Relativity (PoR) asserts the impossibility of detecting inertial motion by any physical means or experiments within the reference frame. The PoR is a common postulate in both the STR and the LPTR, and it is experimentally well supported. In empty space, only the relative motion between two inertial frames has any physical sense, since there is not even a notional reference to measure the motion of a frame, other than another equivalent frame. The PoR, as well as the nature of certain electromagnetic phenomena in relative motion, were the motivating factors of Einstein's theory. In the STR, it is the clock in the 'other frame' that goes slower due to time dilation, because the observer is always at rest in his own reference frame.

It might be appropriate to warn ourselves to certain potential aberrations in such a theory, from the point of view of physical common sense. I point out this not as an inconsistency of the theory, but only as a cause of suspicion that requires further scrutiny.

A typical scenario of relative motion is a flock of birds or an aircraft moving at the velocity v relative to an inertial observer. The STR asserts that the clocks carried in the aircraft, or even the internal clocks of the birds, would run slower than the clocks at rest with the observer. However, suppose the observer sees that the birds are flying backwards! This creates a difficulty because we know that it is in the class of impossible motions. It is only when the observer is moving faster than the birds or the aircraft that he should see them flying backwards. Then, even though there is no precise knowledge of the individual speeds, common sense informs us that the clocks with the observer are moving faster than the clocks in the aircraft. Shouldn't the clocks with the observer have more time dilation than the ones in the impossible backward relative motion? But, Einstein's theory doesn't care! If v is the relative velocity of motion, *physically impossible or not*, clocks must be running slower by the factor $v^2/2c^2$.

I will not discuss the apparent clash of a successful theory with the common sense of the layperson here. But, we will face this situation again when we discuss real experiments of clock comparison. Only experiments can answer subtle questions of physics reliably. And we shall see clear answers to our questions.

3.4 Motion and Its Physical Reference

We briefly discuss the important issue of the necessity of a reference for gauging motion. This is a problem that was discussed even by ancient natural philosophers. It is clear that without material references, relative motion cannot be sensed. In fact, this is spelt out strongly by the Principle of Relativity. The impossibility of sensing uniform motion is obvious in empty space. What was surprising was that even with the 'existence' of a hypothetical medium like the ether that supported the propagation of light, inertial motion could not be sensed. The theories of the relativity of motion, in empty space (STR) as well as in the ether (LPTR), have the strong constraint of the PoR, of not being able to sense inertial motion or the true state of rest. Neither the empty space nor the hypothetical ether is a useful reference for gauging motion.

However, there is a genuine reference for gauging true motion, known from the ancient times, but realized properly only much after the formulation and completion of all our fundamental theories. That is to rely on the average distribution of matter and radiation in our enormous Universe as the single reference for all motion. This is the generalization of the old navigational reference of 'distant stars' that provided an unchanging and stable reference grid for assessing motion.

The existence of a background radiation, a relic from the hot big bang origin and the evolution of the Universe, is particularly useful. The Cosmic Microwave Background Radiation (CMBR), has a black body spectrum with an equivalent temperature $T_0 \simeq 2.726$ K at present. The radiation, peaking at the microwave wavelengths around 2 mm (frequency 160 GHz), is very uniform and remarkably isotropic (Brush, 1992). But the isotropy is only for an observer who is at rest relative to the cosmic frame – the reference frame defined

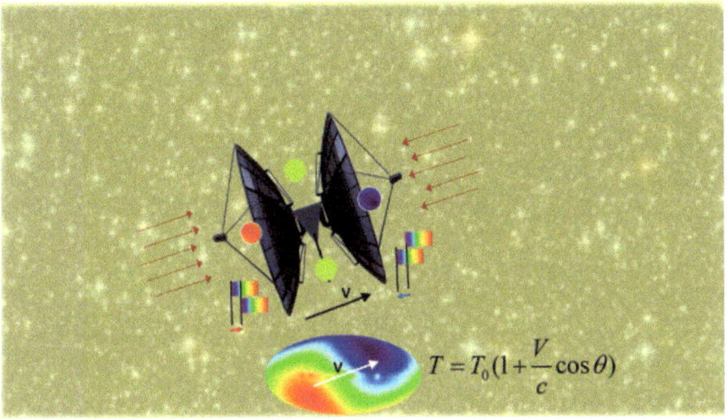

Figure 3.4 The Cosmic Microwave Background Radiation is a relic of the hot origin and evolution of the Universe to the present state, by expansion and cooling. If the isotropic radiation is detected with antennae in a frame moving at the velocity v relative to the cosmic frame, its spectrum will be Doppler shifted towards higher frequencies (blueshift) in the direction of motion, and towards lower frequencies (redshift) in the opposite direction; there is no shift in the perpendicular direction (indicated by the coloured discs). The angle dependent spectral shift is $f' = f(1 + (v/c)\cos\theta)$, which reflects similarly in the effective temperature of the CMBR. This dipole anisotropy is the accurate measure of the absolute velocity in the cosmic frame.

by the large-scale matter distribution in the Universe. If an observer is moving relative to the cosmic frame, the whole spectrum of the radiation (detectable in the range of 10–1000 GHz) received by a directional antenna will be shifted in frequency due to the familiar Doppler shift,

$$f' = f\,(1 + \hat{r}\cdot\vec{v}/c) = f\left(1 + \frac{v}{c}\cos\theta\right) \qquad (3.12)$$

where $\hat{r}$ is the direction of the antenna beam (the radiation is detectable in the range of 1–1000 GHz). This results in a corresponding angle-dependent shift in the equivalent temperature of the radiation, $T' = T_0(1 + \frac{v}{c}\cos\theta)$, called the 'dipole anisotropy (Fig. 3.4). The easily measurable dipole anisotropy of the radiation provides an accurate measure of the velocity of motion, $\vec{v}$. Therefore, no observer who is moving inertially relative to the CMBR can claim to be at rest! That would be inconsistent with the measurement of the Doppler anisotropy of the radiation (Fig. 3.5). Further, the average temperature T_0 of the CMBR decreases very slowly

Figure 3.5 Measures of absolute motion and absolute time in the real Universe. Absolute rest in the cosmic frame (CF) is defined by the stationary distribution of cosmic matter (1a). The slowly changing temperature of the CMBR provides absolute time as well, synchronized everywhere in the Universe to great accuracy (1b). The static matter at true rest becomes a 'flow', or a *matter current*, when there is motion relative to the CF (2a). This is reflected accurately in the Doppler anisotropy of the CMBR (2b).

monotonically and serves as a notional absolute universal clock, synchronized everywhere in the Universe to an accuracy better than a part per million. *Thus, we have an explicit physical realization of both absolute space and absolute time in modern physics and cosmology.*

However, the mere existence of a global reference for motion and time is not the crucial factor for the physics of relativity. The state of absolute rest is also indicated by the stationary nature of the large-scale matter distribution in the Universe. Any real motion breaks this state, and then there is a coherent flow of the cosmic matter, opposite to the motion. *The flow (current) of matter is also the current of the sources of gravity.* This is the crucial fact that makes a fundamental difference to physics. It will turn out that the full import of the cosmic reference frame is manifest in physics through the gravity of the matter and energy in the Universe. The new theory of relativity that I developed based on this necessary gravitational paradigm – Cosmic Relativity – finally grasps the true basis for dynamics and relativity, because it has the factual gravitational Universe as its foundation (Unnikrishnan, 2009). The relation between time and cosmic gravity, as revealed in Cosmic

Relativity, is the central theme of this book, about which we will discuss every relevant issue.

3.5 Simultaneity

When we talk about comparing clocks and their rates, there should be clarity about the methodological aspects of clock comparison. There are two aspects to be considered in the 'relativity' of time; one is the synchronization of two clocks and the other, the comparison of their rates. The major realization on both counts is that *clocks (and time) can be compared without ambiguity of conventions, theoretical assumptions, etc., only in proximity, at the same spatial location*. The synchronization of clocks is intimately connected to the notion of simultaneity: the simultaneous occurrence of two physical events as judged by a single observer in one location.

If the two events are at the same location, not necessarily at the location of the observer, there is no complication in determining whether they are simultaneous. Of course, if both occur at the same location as the observer's position, then not only the order of the events, but also their temporal identity relative to a local clock are obvious. If the events are at the same location, but separated from the observer, there is an unknown difference of time relative to the observer's clock. A messenger for the signals of the events, like light or sound, is required to perceive the events. However, this does not affect the judgment regarding their simultaneity. The full subtlety arises when we want to judge the simultaneity of two events occurring at different locations that are separated from the position of the observer.

The simultaneity of events was conventionally understood intuitively, based on their order in one's perception. This intuitive understanding does not explicitly refer to a clock or time; one event is the temporal reference for the other and their roles are symmetric. It might seem that a universal reference of time is not required for judging the simultaneity of two events. This is true only when the two events happen at the same spatial location. It is easy to see that if two events happen at the same time, hence simultaneously, at the same location, according to some observer (reference frame), then those events are simultaneous to all observers, independent of theories and assumptions. If they happen at different spatial locations, then their simultaneity is ambiguous without referring to a universal clock that shows synchronized time at both locations. In the absence of such clocks, one needs to understand the 'relativity of simultaneity'.

The relativity of simultaneity was first discussed by H. Poincaré in the context of the coordinate transformations between two inertial observers in relative motion (Poincaré, 1898, 1900). This was then closely linked to the synchronization of spatially separated clocks. He clarified that the Galilean nature of the finite velocity of light renders the concept of absolute simultaneity of spatially separated events as ambiguous, highlighting the role of an underlying convention in understanding simultaneity. Einstein discussed the notion of simultaneity in great detail as the fundamental notion in the construction of the Special Theory of Relativity. However, Einstein came to the same conclusion as Poincaré's, of the relativity of simultaneity, while maintaining that the velocity of light is the same invariant constant in all frames! While the lack of universal simultaneity of separated events is a fact, its logical and physical basis in modern discussions is muddled, as we will see later. In particular, we will closely study how both Poincaré and Einstein arrived at similar conclusions on the relativity of simultaneity, in spite of their antipodal assumptions about the propagation of light. The philosopher Henri Bergson's insightful exposition of the central issue will figure prominently in our discussion (Bergson, 1922). This will empower us to make certain crucial conclusions about the correct theory of relativity and physical time.

3.6 Comparing Clocks

What do we mean by 'comparing clocks'? We can compare their reading and also their rates, during the comparison. To gain clarity, let us first discuss the case when the clocks are in the same location, in physical proximity. The readings of the clocks are arbitrary, as set by the user. This initial lack of synchronization is not physically significant, since they can be synchronized before the comparison starts, or the difference may simply be noted, as a constant offset. Once that is done, their rates can be compared directly, by noting the progress of time. Both are valid times and their rate and stability will depend on the details and the quality of their construction.

This arbitrariness cannot be allowed in the clocks used as good measures of time. Hence, there is the concept of standard clocks, following a standard unified definition of the 'second' or the unit of time, as closely and stably as possible. The goodness of a particular (or a particular kind of) clock is established by comparing two identically constructed clocks for a certain duration of time – a natural and 'intrinsic' measure of goodness is the duration they can remain synchronous with each other, or how much they drift away from each other in a fixed duration. For quartz clocks, the mutual drift may be in the range of a microsecond in a second to an hour, and for the Rubidium and Cesium atomic clocks based on microwave transitions, the drift is only a few nanoseconds in an hour to several days. With the laser cooling technology, combined with the faster internal rates available in optical transitions, atomic clocks have been improved significantly. The best laboratory standard clocks maintain their time so well that their mutual drift is only about a nanosecond in 10 years. This may be compared to the relative gravitational time dilation of a nanosecond in a few months due to the gravity of the Earth, of two such clocks separated by 1 m vertically.

How does one ensure that the rate of a clock is 'uniform' and constant? If all kinds of clocks slowly 'accelerate' alike in their rate, there is no way to know this, because measuring the rate of one clock requires a comparison with another. We can only use the physical input that has gone into making a device as a clock to trust in its regularity. For the celestial clock, the large mass and the near constancy of the Earth's angular momentum determine its regularity. For atomic clocks, the laws of electrodynamics and quantum mechanics, as well as the constancy of the fundamental constants that determine the energy levels of the atoms, ensure the regularity.

3.7 Clock Synchronization

It is clear that comparing and synchronizing the times shown by two clocks is easy and conceptually straightforward when the clocks

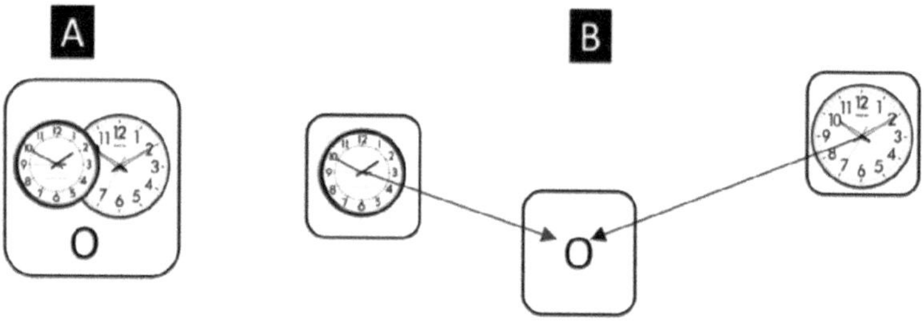

Figure 3.6 Comparing and synchronizing two clocks: (A) Clock comparison at the same location by the local observer O is direct, not requiring the propagation of a signal. (B) Comparison of two spatially separated clocks by an observer O at a third location. This requires a messenger signal, like light.

are in spatial proximity (Fig. 3.6). However, synchronizing clocks at different places is one of the toughest conceptual and technical problems faced by the scientists and engineers entrusted with time-keeping all over the world. For centuries, the most precise reference clock they had access to was the heavy rotating Earth, coupled with the long record of precision astronomy. But, the earth-clock provides only a stable 'rate' of time, and not a reference to the local time. How do we fix the 'zero' of time at each place, say in Mumbai or Paris? We can arbitrarily fix a place as the standard time keeper, provided the nations agree. Then, the 'standard time' needs to be *transferred* around the globe. The apparent motion of the Sun provides an automatic transfer. The solar time – the time at which the Sun is at the local zenith – could then be called the repeatable '12 o'clock local time'. This is an ancient clock that is not accurate or stable enough.

The protocol and the procedure for time transfer and synchronization constitute a difficult task. The difficulty is not merely one of practical implementation; it is conceptually subtle. The technology of the 20th century brought new solutions and challenges. The radio communication aided the method of synchronization with electromagnetic waves over large distances, assuming the invariant constant speed of light. The development of quartz oscillators improved laboratory time-keeping beyond the astronomical clocks. Then, the development of the first atomic standard for clocks, based on the hyperfine microwave transition in the lowest energy state of the cesium atom, by L. Essen and J. Perry in the UK, pioneered the atomic clock era that continues to invent new clock standards. The precision and stability of atomic clocks now exceed a part in 10^{17}. That corresponds to less than a second lapse during the age of the Earth.

Then the question how to synchronize distant clocks with a precision comparable to their intrinsic capabilities arises afresh. Clocks can be synchronized in two ways: it can be done with electromagnetic signals, or the clocks can first be synchronized at proximity by direct comparison and then moved to their intended locations. The second method is less practical, but it is important in the discussion of certain crucial experiments on time dilation. In the first method, one needs a reliable theory of the propagation of electromagnetic signals in free space. (We do not discuss the obvious technical point that we also need to know the details of the propagation in real environments, like the atmosphere, optical fibres, etc., which involve several technical and practical issues.) In the second method, we need a theory of the modification of time in moving clocks. It so happened that the theory of time of *moving* clocks and their synchronization developed from the considerations of the propagation of light, and the two are inextricably mixed in the theories of relativity. However, they appear in distinctly different ways in the pre-1905 Lorentz–Poincaré theory and in Einstein's Special Theory of Relativity.

3.7.1 Space-Time Diagrams

To understand the process of the comparison and synchronization of clocks, as well as the comparison of spatially separated physical events, a graphical representation of events indicating their location and the time of occurrence is helpful. Everybody is familiar with the representation of motion in space as a function of time. When one draws the graph of the progress of the motion of a projectile, one is drawing a kind of space-time diagram: a diagram of spatial motion, showing the height reached by the projectile as a function of time. Conventionally such a high school graph takes time on the 'x-axis' and the spatial distance or height, etc., on the 'y-axis'. The parabola of

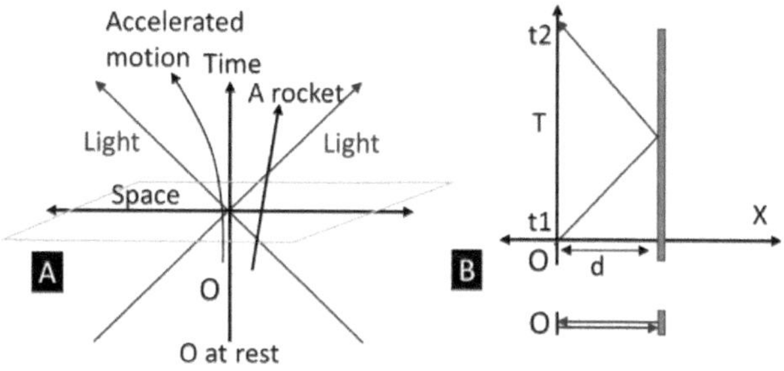

Figure 3.7 Space-Time diagrams. (A) General diagram depicting light propagation along $+x$ and $-x$ at speed c, and a projectile in motion at velocity v<c. Observer O is at rest. The curved trajectory indicates acceleration. (B) Space-time diagram of a light signal starting at time t1 and reaching back at the same location at t2, after reflection in a mirror that is stationary at distance d relative to O.

the space versus time diagram looks similar to the factual trajectory in space, because the horizontal distance travelled by the projectile is directly proportional to time; hence, the 'x-space' and time are identical. This example reinforces our understanding that uniform motion itself serves as a clock, relative to which other motions are measured. When H. Minkowski formulated the STR in terms of the mathematics of the space-time interval, $ds^2 = -c^2dt^2 + dx^2 + dy^2 + dz^2$, the sketch of events in space and time took a more suggestive form (Minkowski, 2012). The axis of time was conventionally taken as the vertical future pointing arrow, and two dimensions of space could be then represented by the x- and y-axes (Fig. 3.7). These drawings assume that the relative speed of light is an invariant constant, equal to c. If we choose the unit of time such that we call ct as time, then both time and space are measured as distances. Hence, we have the units of distance like the light year.

Since every observer in the STR has the same right for claiming that his position is stationary, there is a space-time diagram for each observer O, where O is at rest and the reference frames of other observers are in different motional states. Standing still at a location with a clock is then 'motion' only along the time-axis; time 'moves uniformly' while standing still at a location, 'carrying' the observer along the time axis. (Remember that 'time' is treated as another dimension and one can 'move' along all the four dimensions.) So, nothing can be stationary, remaining at a point, in a Minkowski space-time diagram because time cannot be stopped. Light travels along the 45-degree line (since units of time and distance are the same – light moves 1 unit of distance in 1 unit of time).

Every inertial observer is in space-time motion along a straight line in a space-time diagram, with the slope of the line (measured from the vertical) related to the velocity, with maximum slope ± 1, for light. A curved trajectory indicates acceleration. We will use simple space-time diagrams for discussing clock comparison, simultaneity, etc.

3.7.2 Synchronizing with Electromagnetic Signals

We start with the method of synchronizing clocks with electromagnetic signals, familiar to everybody through the setting of time by radio. The delay in propagation, amounting to tens of milliseconds across the globe, is relatively insignificant for ordinary clock use, but very large for precision use. Even after correcting for the delay in propagation due to the distance, the accurate synchronization crucially depends on the nature of the propagation of light – whether it is Galilean or whether it is Einsteinian. In the Galilean propagation, the speed depends on the speed of the receiver, whereas the Einsteinian light has a constant speed *c*, independent of the speed of the receiver. Both kinds of propagation of the waves do not depend on the speed of the sources, as in the case of sound.

To synchronize a clock C1 to a reference clock C0, an EM signal is sent from C0 to C1 at a definite instant t_s on the clock C0. When C1 receives the signal, its reading is adjusted to be a preassigned value, say '$t0$', and the signal is transmitted back without any delay. When the clock C0 receives back the signal at the instant t_r, it is adjusted to also read '$t0$' at the exact midpoint of the time between sending and receiving the signal, $t0 = t_s + (t_r - t_s)/2$ (Fig. 3.8B). Then, both the clocks are assumed to read the same time – they are synchronized. If the clocks are relatively stationary, synchronizing once is enough. If there is a relative motion between the clocks, then repeated communication is required for meaningful comparison. *The crucial assumption in the conventional synchronization that uses Einstein's STR is that the relative speed of light is the same*

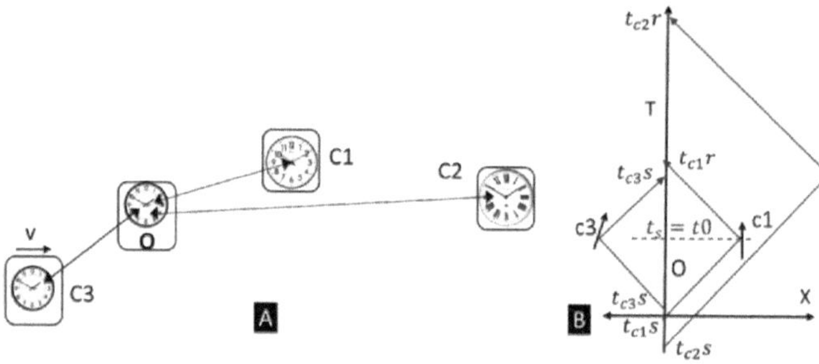

Figure 3.8 Clock comparison and synchronization by an observer O using electromagnetic signals. (A) The distant clocks (C1, C2 and C3) may be relatively stationary or moving as in situations of the GPS navigation (C3 in the figure). (B) The space-time diagram of such comparisons. Signals are sent to the clocks 'n' at definite instants ($t_{cn}s$, etc.) and received from the clocks at instants $t_{cn}r$. The sync-time $t_s = (t_{cn}s + t_{cn}r)/2$ is shown for the clock C1 with a dotted line marked $t_s = t0$.

constant for all inertial observers. With this assumption, light takes the same duration both ways, from C0 to C1 and back, from C1 to C0, independent of whether C0 and C1 have any common inertial motion.

Both the process of synchronization and its verification use the same physical messenger of light. Therefore, the synchronization of two spatially separated clocks is contingent on the hypothesis on the nature of propagation of light. Clock synchronization is dependent on the convention used, though self-consistent. In the present times, the STR and its light hypothesis are the basis for clock synchronization. Since the clocks are at different positions, their synchronization is a matter of faith in the theoretical assumptions and it is not verifiable independently of these assumptions. To stress this point, I repeat that there is no way to find out whether two clocks are 'really' synchronized and reading the same time, if they were synchronized using the standard 'light protocol'.

The problem is that it is not known from any experiments whether the one-way propagation of light is Galilean or Einsteinian! Our knowledge about the propagation of light is from the high precision experiments on the two-way propagation of light, like the M-M experiment and its many variations. All experiments that claim a test of the one-way propagation of light also use spatially separated clocks for such measurements. But since the synchronization of the clocks depends on whether light is Galilean or Einsteinian, the experiment is rendered logically circular! If the synchronization is done assuming the Galilean propagation of light, then the test with such clocks returns the result that the propagation is Galilean. On the other hand, if the clocks are set assuming the Einsteinian light, the result on the propagation will turn out to be Einsteinian. Thus, both postulates are consistent with what is observable in such experimental tests. This is why Einstein's well-known postulate of the absolute constancy of the relative velocity of light (the light hypothesis) has remained as a postulate even after a century. The widespread belief that the M-M experiment or any other experiment prove the postulate of the invariant relative velocity of light, is the result of the ignorance about the physics of synchronization of clocks and also about the inefficacy of the two-way experiments to provide a conclusion on the one-way propagation. We have already discussed this in the Section 3.2.

We can now proceed to examine the protocol for synchronizing two clocks A and B in more detail, to understand how it is inextricably linked to the Lorentz Transformation for time and the propagation of light. The first step is the signal from clock A to clock B at time t_{A1} in clock A. When this is received at clock B, its time is set as $t_{B0} = 0$ and the signal is sent back at this instant. When A receives the signal at t_{A2}, its time is set as $t_{A0} = 0 = t_{A1} + (t_{A2} - t_{A1})/2 = (t_{A2} + t_{A1})/2$. Now the clocks are synchronized. This is the protocol discussed by Poincaré and later, by Einstein (Poincaré, 1904; Einstein, 1905, 1920). But, their assumptions on the propagation of the signal were different. For Poincaré, the relative speed of light was Galilean, in the ether medium. However, since we cannot know any inertial motion relative to the ether (PoR), there is no way to gauge the relative speed of light. Hence, *the synchronization of clocks employing light signals is only a consistent and practical protocol under a convention.* It has no implication of the true synchronization of time. In sharp contrast, Einstein asserted the absolute constancy of the relative velocity of light as a physical truth. Hence, the light signal synchronization of every clock at relative rest is assumed to be the true synchronization in the STR.

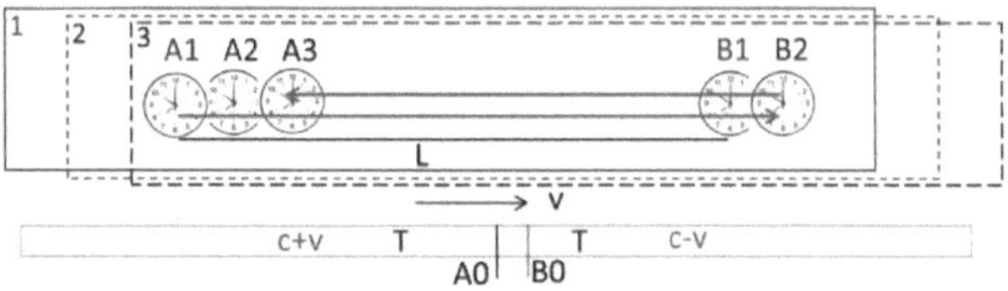

Figure 3.9 Synchronization by light signals in a frame that moves at velocity v. The successive positions of the clocks at the locations A and B are labelled A1, B1, etc. Observers at A and B are *unaware of their motion and assume that light signals take equal times in their up and down trips,* A to B and back. Clock B marks zero when it receives the light pulse and A marks zero at the midpoint of the total duration $2T$. In reality, the clock moves B1 to B2 while the signal propagates, taking more time to reach (long right arrow), and it is received back when the A-clock is at A3, taking less time in transit. Then the local time set at B is behind the time at A, while the observers at B *believe* that they have the same time as at A. If the separation between the clocks is L, the total duration $2T = 2L/c$ is divided in the ratio $t : 2T - t = (c + v) : (c - v)$. Thus, $B0$ is not the same as $A0$; it is $\delta t = Lv/c^2$ later than $A0$.

As a member of the Bureau des Longitudes, Poincaré had studied the problem of the global synchronization of time, before making decisive founding contributions to the principle and theory of relativity (Fig. 3.9). He had analysed the situation very clearly and wrote (Poincaré, 1904),

> I suppose that observers placed in different points set their watches by means of optical signals; that they try to correct these signals by the transmission time, but that, *unaware of their translational motion and thus believing that the signals travel at the same speed in both directions,* they content themselves with crossing the observations, by sending one signal from A to B, then another from B to A.

I have italicized the essence of the interpretation. Hence, it is clear that Lorentz's local time is the mismatch between two clocks that are synchronized at relative rest, *believing that they were at absolute rest, unaware of the motion of the frame at the velocity v*. If two clocks, 'synchronized' thus, are used to measure the relative velocity of light, the fact that light takes extra time at a smaller relative velocity would not be detected. It is obvious that the measurement of the velocity of light done using such clocks, synchronized assuming the equal velocity both ways, when it is really $c - v$ and $c + v$, will always return the result that the velocity is an invariant constant, independent of the velocity of the frame. One cannot establish the invariance of the relative velocity of light using spatially separated clocks that are synchronized using electromagnetic signals.

In the STR, the time taken to propagate the distance L between the clocks at relative rest is $T_E = L/c$, in both directions, irrespective of whether the clocks have a common velocity or not. In contrast, in a theory of relativity with an absolute frame (abbreviated as ATR, which includes relativity theories with a single privileged frame like the ether or the Universe), with the Galilean propagation of light, $T = L/(c \pm v)$, where v is the velocity of the common frame in which the clocks are at relative rest. (The minus sign applies when the signal to the clock B is in the same direction as the common velocity through the absolute frame.) The difference in the duration of propagation is

$$\delta T = T - T_E = \frac{L}{c \pm v} - \frac{L}{c} \simeq \mp \frac{Lv}{c^2} \tag{3.13}$$

Therefore, the duration of propagation is different in the two directions. Then, the synchronization by the light protocol will be *factually asynchronous* by the amount Lv/c^2, which depends on the separation between the clocks as well as on their common velocity. Following the protocol in detail, clock B marks its 'zero' when it receives a signal sent from clock A at t_{A1}, after a propagation delay of $L/\gamma(c - v)$. Now I am being more precise by including the length contraction factor. Clock A marks its 'zero' at the time $t_{A1} + T_p/2$, after it receives the return signal after a total propagation delay T_p. But, the two-way propagation delay now is not $2L/c$; it is

$$T_p = \frac{L}{\gamma(c - v)} + \frac{L}{\gamma(c + v)} = \frac{2L}{\gamma c(1 - v^2/c^2)} \tag{3.14}$$

(Note that this is the same 'two-way' duration in the M-M experiment, Section 3.2.) The difference between the two clocks $(T_B - T_A)$ is

$$\delta T = \frac{L/\gamma}{c \pm v} - \frac{L/\gamma}{c(1 - v^2/c^2)} = \frac{L(1 \mp v/c) - L}{c\sqrt{1 - v^2/c^2}} = \mp \frac{\gamma Lv}{c^2} \tag{3.15}$$

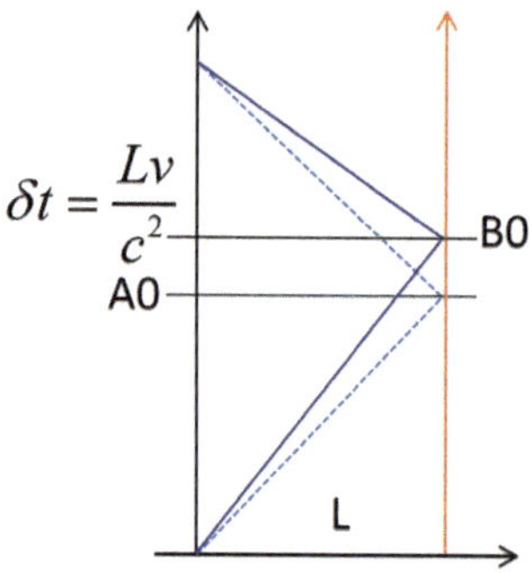

Figure 3.10 Clock synchronization in a space-time diagram. The dashed diagonal lines indicate the belief about the propagation of light and the blue lines indicate the Galilean reality, in a moving frame. The local time at B ('0' of B) is lagging by Lv/c^2 relative to A0, which is at a distance L away. But, A (or B) *cannot* know this gap in synchronization.

Hence, the synchronization of clocks separated by distance L at relative rest, with the Galilean light and the standard protocol, in a frame moving at the velocity v, results in a factual gap in synchronization (sync-gap) of $\delta T = \mp \gamma Lv/c^2$. Clock B is really lagging behind clock A by $\delta t = Lv/c^2$. That is, the 'zero' of clock A is factually earlier than the 'zero' of clock B (Fig. 3.10). When t is the real time, what clock B reads is $t - Lv/c^2$; this is the real 'local time' at B. *Because of the PoR, we remain unaware of this sync-gap between the spatially separated clocks.* The first-order term $-Lv/c^2$ in the LT is the theoretical compensation for this Galilean gap, which then retains the first-order invariance of electrodynamics. The second-order term in the Lorentz Transformations, $\gamma = 1/\sqrt{1 - v^2/c^2}$ is common to both the clocks and it does not affect the synchronization.

In terms of a concrete example with realistic numbers, if we attempt to synchronize two clocks separated by about 25000 km (as in the GPS satellites) while the solar system is moving through the Universe at the velocity v of about 400 km/s, the term Lv/c^2 could be about 10 μs. Thus, when the atomic clocks on the Earth read T, the 'precision clocks' in the satellites, which are 'synchronized' with light (at the relative velocity $c - v$), will be lagging by a whopping 10 μs! Remember that the atomic clocks in use are more than a million times accurate than this sync-gap. How come we do not observe this huge discrepancy? The reason is simple. *We can know of the readings of the satellite clocks only if we bring back the signal from those clocks to our location.* When the satellite clocks send their time codes towards the Earth, the situation is reversed; the signals reach 10 μs earlier because now the relative velocity in the return trip is $c + v$. *The large 10 μs gap in the clock synchronization is then rendered invisible in the two-way protocol.* The physical reality remains superficially hidden, increasing the faith in the wrong theory that asserts the constancy of the relative speed of light.

However, this is precisely the kind of 'asynchronous-synchronization' of clocks that ensures the apparent invariance of the relative velocity of light, and the invariance of Maxwell's equations. Thus, we have completely decoded the Lorentz Transformation for time,

$$t' = \gamma(t - Lv/c^2) \tag{3.16}$$

Therefore, it is clear that the LT of time is designed to mask the gap in the clock synchronization using the Galilean light. That is the reason why light *appears* to have an invariant relative velocity under the Lorentz Transformations. Time in clocks and the propagation of light are intertwined in a circle of physics and logic. *The constancy of the relative velocity of light is an apparition, and not a real physical feature of light.* If clocks 'synchronized' using light are used in an experiment to measure the speed of light, the experiment cannot reveal the Galilean nature of light because of the sync-gap caused by the very Galilean nature of light! While that is what is happening in a real experiment, we incorrectly interpret it as due to the constancy of the relative speed of light, because we forget that the synchronization of the clocks was done using light itself. Light has fooled our clocks and time to hide the ordinary nature of its fast journeys. And the deceit has worked for over a century, but no more...

We have discussed the intertwined relation between the propagation of light and the synchronization of clocks in great detail, from many angles, because of its paramount importance in dealing with the time of clocks at different spatial locations. The comparison of time at different locations is faced in our everyday activities and in all spheres of life, and we cannot afford to be misled by the unfounded beliefs in the factually incorrect theories of light and motion.

3.7.3 Clock Synchronization by Slow Transport

The other obvious possibility for ensuring the synchronization of distant clocks is to first perform the synchronization operation *in proximity* and then move the clocks to their intended locations. However, since this depends on what happens to the rate of the clocks in motion, which in turn depends on the physical theory one assumes, one has to find a way of avoiding (or accounting for) the dependence on the velocity of the clocks during the transportation. Since it is known from the elementary theory of relativity and from experimental evidence that the dependence is quadratic on the velocity, $(1 - v^2/c^2)^{1/2} \simeq 1 - v^2/2c^2$, one might think that pre-synchronized clocks could be moved to their required locations keeping the motional time dilation negligible, if the speed of the movement is very small. Even otherwise, since we know the formula for the time dilation, we could correct for whatever change of time that occurs during the clock transport, in principle, if we keep track of the trajectory of transport. However, this does not solve the problem of true synchronization of time across spatially separated clocks! The catch is clear from Fig. 3.11. The clocks are relatively moved at a known relative velocity. But, we cannot know whether there is a common prior velocity v_f at which both the clocks are moving, consistent with the Principle of Relativity, unless an external absolute reference is used to measure all motions.

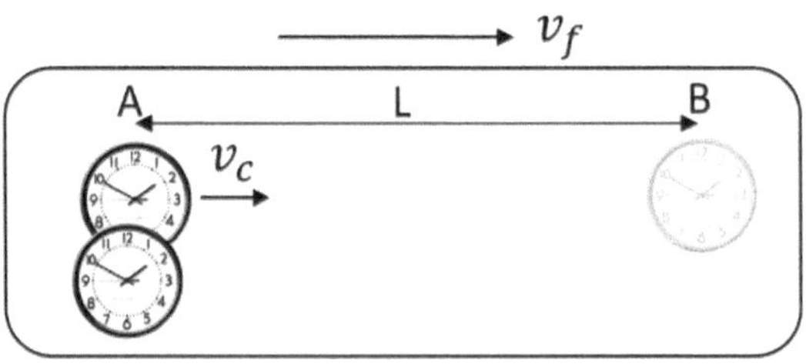

Figure 3.11 Clock synchronization by the slow transport of one of the pre-synchronized clocks between the locations A and B, at the velocity v_c. No electromagnetic signal between the spatially separated clocks is involved.

For a clock transported at the small velocity v_c in the frame moving at the velocity v_f, the time dilation over the duration of transport T_B is

$$\Delta T_B \simeq -T_B \frac{(v_f + v_c)^2}{2c^2} \simeq -T_B \left(\frac{v_f^2}{2c^2} + \frac{v_c^2}{2c^2} + \frac{v_f v_c}{c^2} \right)$$
$$= -T_B \frac{v_f^2 + v_c^2}{2c^2} - \frac{L v_f}{c^2} \tag{3.17}$$

The last term is obtained from the relation $v_c T_B = L$. The term $T_B v_f^2/2c^2$ is common to both the clocks. The term $T_B v_c^2/2c^2$ can be fully corrected for, because v_c is known from the trajectory of transport. Also, this term can be made negligible by choosing very small v_c, when practical. However, there is an *unknowable* first-order time dilation between the two clocks, of $(L/c)(v_f/c)$, making the clock B to run *behind* A, if synchronized with this 'slow transfer' protocol. Again, we are stuck with a sync-gap of $-Lv/c^2$, similar to the sync-gap when the synchronization was attempted by light propagation. *True synchronization that is independent of the background theory is impossible.* The only exception to this involves the universal absolute time provided by the evolution of the Universe, as we shall see later.

If we use such a clock to measure the propagation duration of light, the synchronization mismatch $-(L/c)(v_f/c)$ exactly masks the Galilean propagation factor $(L/c)(v_f/c)$. *The Galilean light takes the extra duration Lv/c^2, but the synchronization protocols leave the clocks with exactly the same duration less than the true synchronization.* Since the measurement of the velocity of light needs synchronized clocks at two locations, one invariably ends up with the false conclusion that the speed of light is an invariant! The true facts about the relativity of clocks and the propagation of light are not what is believed and stated widely. Our detailed analysis shows plainly, and in precise quantitative terms, where we have gone astray in our judgment of the facts.

We will return to the issue of clock synchronization and comparison when we discuss the Navigational Satellite Systems like the GPS.

3.8 What Causes Time Dilation?

The mathematical expression for the elemental time dilation of a moving clock in the STR is very simply stated; if the clock moves the relative distance dx in time dt (the relative velocity is then $v = dx/dt$), it will read the time $d\tau = dt\sqrt{1 - v^2/c^2} \approx dt(1 - v^2/2c^2)$ instead of dt. This can also be formally written as

$$d\tau^2 = dt^2 - dx^2/c^2 \tag{3.18}$$

Any spatial distance covered by the clock, relative to an inertial frame, results in a slower rate of time. This gives a mysterious aura to the spatial distance covered in relative motion, as if space had the ability to wrest out time from clocks.

At this stage, we should take a contemplative rest and deal with the *important question of what causes the physical time dilation in a moving clock.* This question is particularly important if we believe that the theoretical framework explaining this easily observable physical phenomenon is the STR, because the sole 'velocity' parameter that determines the rate of a clock is the 'relative velocity', with reference to any inertial frame. Hence, this parameter does not have a fixed value, unlike the 'relative velocity' in a theory with a master frame (like the Universe). The phenomenon of time dilation involves the comparison of two individual clocks. Since they operate independently of each other, *the relative time dilation is the difference between the time marked by each clock.* However, if time dilation is a physical phenomenon, each clock will have its own modification of the rate, which will reflect in the relative time when a comparison is made. Presumably, and quite reasonably, the modification of the rate of each clock depends on its individual motion history. But, between two clocks in two inertial frames, the partition of the relative velocity v into two individual velocities v_1 and v_2 is arbitrary, and there is an infinity of valid partitions, the only constraint being $v = v_1 - v_2$! This is a serious conceptual problem – what is the contribution of each clock to the time dilation? The STR does not have an answer, because it deals with only the derived quantity, the relative velocity. An ATR has a clear answer, since each clock has its time dilation that depends on its speed relative to the master frame. Then the relative time dilation is just the difference between the individual time dilations.

Since there is no 'true' motion in the STR, and the relative velocity depends on the choice of the frame, the apparent motion of an individual clock cannot be the cause of the time dilation in the STR. Even if we attribute the cause as 'motion', we face the problem that a physical change in the material clock has happened without any identifiable interaction, whereas all observable physical changes in material systems are always due to one of the interactions, in the experience of physics.

We have to admit then that there is no physical cause for the motional time dilation of the STR. An effect without a cause? Could such a thing even happen? In any case, one could make the categorical assertion that the time dilation in the STR has nothing to do with gravity, the only interaction capable of universally affecting the rate of clocks, because the STR explicitly assumes empty space and no gravity.

I can disclose a very surprising fact, the details and reasons of which we will study later (Chapter 7). Navigational satellite systems like the GPS are supposed to be using the STR as the guiding and working theory. However, formulae used to calculate the corrections due to the motional time dilation in the GPS are explicitly from a theory of relativity with an absolute frame, and not from the STR! It is not the relative velocity that determines the motional time dilation in clocks. In all cases known, the correct motional time dilation of clocks is obtained by using the expressions in a theory with an absolute frame as well as the Galilean propagation of light.

3.9 Imaginary Time and Real Space

The elementary Pythagoras formula $a^2 + b^2 = c^2$ relating the lengths of the sides of the 'ABC' right triangle is the basis for finding the distance between points at close proximity in a space of any dimensions. Then the distance between the vertices A and C is $c = \sqrt{a^2 + b^2}$. In three dimensions, the distance between two nearby points separated by the orthogonal coordinates x, y, z and

$x + dx, y + dy, z + dz$, is $dl = \sqrt{dx^2 + dy^2 + dz^2}$. How does a 'dimension' of time fit into this scheme of distances?

The concept of space-time as a homogeneous physical entity has become an integral part of physics, but the real difference between space and time is hidden in their mathematical representation, unnoticeable to most. A general impression is created about the reality of a four-dimensional world, where time is one coordinate and the other three are usual spatial coordinates. However, the squared distances in this peculiar world are not the sum of the squares of the coordinate intervals – it is the *difference* of the squares of the spatial distance and the temporal duration.

The three coordinates of space are combined with time to form a four-dimensional entity, called space-time. This is a free creation of the mathematical mind. There is no natural way of treating the 4 numbers as describing a single geometrical point in a four-dimensional physical world. This is because the 'distance' ds between two nearby points in space-time is $ds = \sqrt{dx^2 + dy^2 + dz^2 - c^2dt^2}$, *with a minus sign for* dt^2, instead of the conventional 'square root of the *sum of squares*' formula. Consequently, the time coordinate has a hidden imaginary content; $-c^2dt^2 = (icdt)^2$. However, this can be made to look like the conventional 'sum of squares' Pythagoras formula by transferring the burden of the negative sign to a constant 'metric', η_{ij}, where $\eta_{11} = \eta_{22} = \eta_{33} = 1, \eta_{44} = -1$, and $\eta_{ij} = 0$ when $i \neq j$. Then

$$ds^2 = \eta_{11}dx^2 + \eta_{22}dy^2 + \eta_{33}dz^2 + \eta_{44}c^2dt^2 \tag{3.19}$$

Now, it looks as if the time coordinate has the same nature as the spatial coordinates. However, the imaginary nature and the physical strangeness remain, because the minus sign is still there, carried by an inseparable metric.

The basis of the special relativistic assertion 'time=space' is treating cdt as a distance. But where does the imaginary character spring from? The easiest way to see the physical interpretation is to consider the propagation of light between two points x, y, z and $x+dx, y+dy, z+dz$ for which the equation $dl = cdt$ applies literally. The spatial distance between the two events, of the emission of light flash at the point x, y, z at time t and its reception at the point $x + dx, y + dy, z + dz$ at $t + dt$, is $dl = \sqrt{dx^2 + dy^2 + dz^2}$. The equation $\sqrt{dx^2 + dy^2 + dz^2} = cdt$ has the simple meaning that in duration dt light travels a distance $dl = cdt = \sqrt{(dx^2 + dy^2 + dz^2)}$. Squaring, the 'light event' has the equation

$$dx^2 + dy^2 + dz^2 - c^2dt^2 \equiv ds^2 = 0 \tag{3.20}$$

The quantity ds is called the 'line element', which is a 'distance' in the four-dimensional space-time. The resemblance to a 'sum of the squares' of coordinate intervals can be mimicked by treating the duration as a coordinate interval equal to $icdt$. That is, by defining $ict = d\tau$, one can write the equation for the propagation of light as $ds^2 \equiv dx^2 + dy^2 + dz^2 + d\tau^2$, which is indistinguishable from the distance (squared) in ordinary geometry. But, the hidden imaginary nature of the coordinate interval cannot be avoided, which reflects in the value of that distance, which is zero!

We have already seen in detail that by agreeing to consider two clocks in a frame moving at velocity v as synchronized when the time of a clock that is at distance dx from another reading time t is set at $t - vdx/c^2$, the fact that light travels at the relative velocity $c - v$ in the frame remains hidden. Further, the length dx' is contracted and the duration dt' is dilated by the factor $1/\gamma = \sqrt{1 - v^2/c^2}$ in the moving frame. Then, writing only the 'x' and 't' coordinates for brevity,

$$\begin{aligned} dx'^2 - c^2dt'^2 &= \gamma^2(dx - vdt)^2 - c^2\gamma^2(dt - vdx/c^2) \\ &= \gamma^2(dx^2 + v^2dt^2 - 2vdxdt) - \gamma^2(c^2dt^2 + v^2dx^2/c^2 - 2vdxdt) \\ &= \gamma^2dx^2(1 - v^2/c^2) - c^2\gamma^2dt^2(1 - v^2/c^2) = dx^2 - c^2dt^2 \end{aligned} \tag{3.21}$$

Thus, the relation $dx^2 + dy^2 + dz^2 - c^2dt^2 = 0$ is preserved in any inertial frame, when using such clocks (which are clocks synchronized with light or by slow transport over the separation dx).

What about two arbitrary physical events at x, y, z and $x + dx$, $y + dy + z + dz$ at time t and $t + dt$? Then, of course, the difference $dl^2 - c^2dt^2$ is not zero. The difference ds^2 can be positive or negative. However, the invariance of this difference depends only on the use of clocks synchronized with the Lorentz local time included, and on the physical length contraction and time dilation; it does not matter whether the interval ds is zero or not. It follows that the quantity

$ds^2 \equiv dx^2 + dy^2 + dz^2 - c^2dt^2$ is invariant in inertial motion at any velocity v, provided we use clocks that include the Lorentz local time $-vx/c^2$.

Since the invariant interval $ds'^2/c^2 = ds^2/c^2 = dt'^2$ when dx', etc., are zero, ds^2/c^2 can be identified with the square of the time interval marked by a clock carried with an observer. The quantity $ds/c \equiv d\tau$ is called the proper time.

3.10 Can I Change Your Time by Moving My Clock?

We get so accustomed to the concepts and terms in physics by repeated use over decades that we tend to forget the real composition of certain notions. This can lead to serious fundamental confusion. It is important to discuss how this 'sociological' aspect has affected the discussions on the modification of time and clocks in motion.

We are well aware of the fact that a quantity like the *relative velocity* involves the velocities of two distinct entities. The relative velocity itself is a derived quantity; it is the ratio of the relative distance and relative time. Such relative physical quantities are obtained by taking the *difference of two quantities that are measured independently*. For concreteness, *the relative time is the difference of the readings of two clocks*, which are independent. Therefore, the relative time can change only by changing either the rate of one of the clocks or by changing both differently. To be explicit about this important point, I emphasize that the relative time is obtained by taking the difference between two independently obtained measurements of time, which could involve two clocks at different locations, or in different reference frames with a relative motion between them, etc.

In the Special Theory of Relativity, a theory in which the relative rate of clocks depends only on the relative speed, we tend to forget that the time dilation predicted by the theory is the difference of the durations recorded in two different clocks that are entirely independent, without any interaction during their motion. The relative time can change only if the individual clocks change their rates differently. Such changes happen independently in the two clocks because, after the initial synchronization at the same location, all that is physically identifiable is the fact that the separation between them is increasing, indicating a relative velocity; there is no physical sense in asking which clock is 'really' moving. The final comparison of elapsed times is possible only by altering the trajectory of one or both of the clocks, to bring them together.

In the usual discussions on the relative time dilation of clocks, as in the twin-paradox, etc., one often sees the statement that the relative time changes in a decisive way when one of the clocks accelerates to alter its trajectory, creating an asymmetry. Nobody ventures a discussion in which both clocks accelerate by different amounts to adjust their paths to come together again for a comparison. Then there is no qualitative asymmetry and one will have to discuss how exactly the accelerations lead to the asymmetric time dilation, quantitatively.

Consider the problem of the relative time dilation of two clocks that separate after synchronization, due to their relative motion. How does one decide theoretically which of the two clocks has the larger time dilation, when they are brought together for a comparison? Take the case where only one of the clocks is decelerated in its trajectory, to reverse its path and bring it back for the comparison of their readings. It is a fact, learned from experiments, that there is indeed a relative time dilation – the initial synchronization of the clocks is altered after the journey. But, we still need to answer how the relative time was altered – in which clock did the modification of time take place? *If the acceleration is the source of the change in the relative time, only the rate of the clock that was accelerated should change and not the rate of the other clock that remained inertial, far away.* We don't want to upset Einstein with another 'spooky action at a distance', do we? However, if only the rate of the accelerated clock can change, we can easily check the amount of change by comparing with nearby clocks that did not accelerate, at least approximately!

The important lesson is that I cannot change the reading in a distant clock by shaking my clock. The relative time can be affected only by the factual changes in individual clocks. But, there is no way in physics to change the rate of a distant clock by accelerating a local clock. Any claim to the contrary is not science. This insight about the relative time – that it is the difference in the readings of two independent clocks – is important for our discussion of the much-debated 'Twin-Paradox'.

3.11 The Twin-Clocks Problem

The Special Theory of Relativity was projected as a theory of space and time, rather than merely as a theory of the modifications of distances and durations, after H. Minkowski's mathematical formulation, in which time was depicted as a fourth dimension. However, the fact that this dimension had the imaginary nature, in the mathematical sense, confined the geometrical theory within the circle of physicists and mathematicians. It was through a lecture by Langevin at the International Congress of Philosophy in Bologna in 1911 that many attendees were jolted into a discussion about the multitude of times in the theory of relativity (Langevin, 1911). There, Langevin mentioned about time dilation of transported clocks in Einstein's theory, and discussed two concrete examples. One was about the comparison of the lifetimes of two samples of radioactive Radium in relative motion, and another was about comparing the ageing of a fast-moving hypothetical space traveller with the ageing on the Earth. Langevin said,

> Imagine a laboratory attached to the Earth, the motion of which can be considered as uniform translation, and in this laboratory there are two perfectly identical samples of radium. What we know about the spontaneous evolution of radioactive materials allows us to say that if these samples are kept in the laboratory, they will lose both their activity the same way over time and their activities remain continuously equal. But then send one of these samples with a sufficiently high velocity and then bring it back to the laboratory; this requires that at least at certain times this sample has undergone accelerations. We can say that on return, its proper time between departure and return is less than the measured time interval between these events by observers attached to the

> laboratory, so that it has less evolved than the other sample and therefore it will be more active than the latter; it will have aged less, having been more agitated.

As the next example of 'lived durations' of the traveller he said,

> For this it is sufficient that our traveller consents to be locked in a projectile that would be launched from the Earth with a velocity sufficiently close to that of light but lower, which is physically possible, while arranging an encounter with, for example, a star that happens after one year of the traveller's life, and which sends him back to the Earth with the same velocity. Returned to the Earth he has aged two years, then he leaves his ark and finds our world two hundred years older, if his velocity remained in the range of only one twenty-thousandth less than the velocity of light. The most established experimental facts of physics allow us to assert that this would actually be so.

This second example evolved into the fable of the 'twin-paradox' of Einstein's relativity (Fig. 3.12). One of the twin brothers goes out in a long and fast space-voyage. When he returns home, he has aged much less than his brother who stayed back. What is the explanation for the unequal ageing? Langevin was aware and careful about the fact that, in the STR, any observer can claim the state of rest and the special relativistic ageing was symmetric, unless some kinematical asymmetry was invoked. He cited the asymmetry of the accelerations of one of them, required to make the return trip for

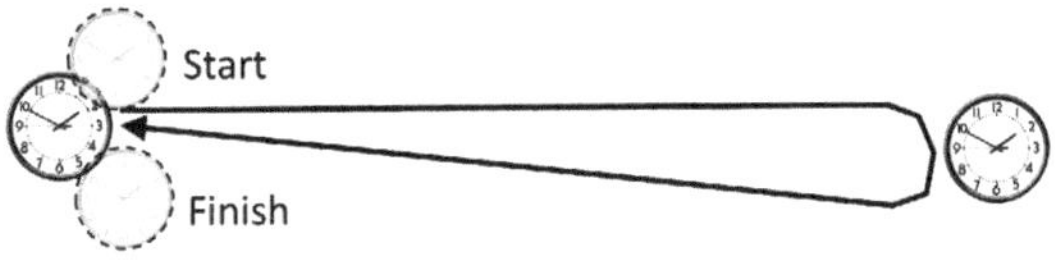

Figure 3.12 The standard scenario for the twin-clocks paradox. There is a relative motion between the two clocks, with a phase of non-inertial motion (acceleration and deceleration) for one of the clocks, when its motion reverses for the return. The clocks are compared at the start and end of the experiment. No electromagnetic signals between the distant clocks are involved.

the eventual comparison at the same place. He stated that the one who had aged the least is the one who had suffered the greatest accelerations. He devised a scheme of light signals for the continuous communication between the traveller and the earth-station and demonstrated the asymmetry as equivalent to the asymmetry of Doppler shifts of the light signals,

> ...the traveler sees the Earth moving away and approaching during times that are equal to one year, while the Earth, which is informed of the acceleration only by the arrival of light waves, sees the traveler moving away during two centuries and returning during two days, i.e. during a time forty thousand times shorter.

However, there was nothing in the physical basis or in the mathematical structure of the STR that could justify the assertion, 'one who had aged the least is the one who had suffered the greatest accelerations'. In the STR, the time dilation depends only on the speed of motion, and not on the acceleration. Those who do not agree with Langevin's unjustified reasoning would naturally interpret this situation as a paradox – hence the twin-paradox, where one person remains on the Earth (A) and his twin brother (B) is the space traveller. One would argue that in the space traveller's rest frame, it is the earth-dweller who is relatively moving and Einstein's theory (through the Lorentz Transformation on time) predicts that the earth-clocks (A) run slower than the clocks of B. There is nothing in the STR which demands a special treatment while B is turning back – that is a small duration compared to years of travel. So, relative to B, it is A who ages slower, whereas relative to A, it is B who ages slower. Since both have equal special relativistic right to claim the state of rest, both are right and the only logical conclusion, without invoking any new theory, is that neither will be slower or faster. Hence, time dilation in the STR must be an unphysical illusion – that is the essence of the twin-paradox and the critique of the Special Theory of Relativity. This problem doesn't arise in the Lorentz–Poincaré theory, where time dilations occur only while moving relative to the ether, or in a theory in which the Universe is the absolute frame – the situation is asymmetric from the start to the end, without accelerations. In any case, the

Special Theory of Relativity does not specify a reason for a clock that underwent a brief period of non-inertial motion to go slower than the one that didn't.

Langevin's 1911 solution remains as one of the most popular answers to the paradox in modern textbooks. Few realize that Einstein had rejected the resolutions entirely within the STR as incorrect already in 1918! In a paper written in the form of a dialogue between a critic and a physical relativist (Einstein, 1918), he first agreed that the paradox could not be solved within the observer-symmetric Special Theory of Relativity; the resolution of the paradox needed a new theory. Einstein's awaited and much delayed answer for the puzzle of the asymmetric time dilation was that the pseudo-gravitational fields experienced during the acceleration phase (during the reversal of the velocity for the return journey) led to the required amount of *gravitational time dilation*, according to his new theory of gravity – the General Theory of Relativity; therefore, there was really no contradiction or paradox.

The traditional debate regarding the twins' ageing was about who aged more after the trip, Peter or Paul, Langevin's original characters. We know empirically that some transported clocks, like the traveller Paul, age less. We also know, from experiments, that some transported clocks age more when they come back after their trip, contradicting Einstein's relativity! *A clock moved from Mumbai to Paris by surface transport, after synchronizing with a reference clock, will actually age faster, and not slower*. To check this, one can accumulate the faster ageing by transporting it further to Washington DC, to California, Tokyo, Hong Kong, Kolkata, and to Mumbai. The clock with the traveller Paul would be about 200 ns *older* than the reference clock at Peter's home, which never left the room in Mumbai. The reason for this cannot be found within the Special Theory of Relativity. But in theories with an absolute master frame, this is easy to understand. We will of course be discussing this in detail later (Section 5.8).

We go back to the question that is usually debated. First, suppose one knew about only Special Relativity. Now we have three characters of the same age, the triplets Peter, Paul and Pauline. Paul and Pauline are in their spaceships that are moving together, side by side. Peter, who is on a large space station, perhaps like a planet, is moving away relative to them. The large velocity of separation is v. Their clocks are synchronized before they separated. Who ages faster? We see that we are unable to answer this question, because we formulated the question simply and correctly! Try answering it and we are immediately stuck, because the STR is a symmetric theory. Paul and Pauline think that Peter should age slower because they see him moving away, while they are at rest in their spaceships. During their elapsed time T seconds, Peter should age by about $T(1 - v^2/2c^2)$, or Peter should be younger by $Tv^2/2c^2$ seconds. In contrast, Peter thinks that Paul and Pauline, who are both moving away, should be the ones who are younger by $Tv^2/2c^2$ seconds. In a theory with a master frame, like the matter-filled Universe, or even the old ether, there is no confusion. The velocity of motion has an absolute reference. Those who move relative to the master frame, age slower. But in the STR, at the end of half the story, a few years of relative motion and separation, there is no definite answer to our question. Then the adherent of the STR will ask, 'then what happened?', looking to break the symmetry.

Shouldn't a theory be able to give the answer at all times? Surely, one of the clocks should be slower than the other, if such a physical effect is really there. Which one?

At least one fact is clear – Paul and Pauline, being side by side in their spaceships, age the same way, with their clocks remaining synchronized. The clocks of the triplets have some readings locally, and whatever they are, a clock's reading in one spacecraft cannot be changed suddenly by firing a rocket in another spacecraft. Nobody would agree to a 'spooky action at a distance' in relativity's clocks.

Pauline decides to fire a rocket briefly to arrange her flight path, to reach and rejoin Peter. So, if at all acceleration can change the rate of a clock, that must affect only the reading of Pauline's clock, and not Paul's or Peter's. Her spaceship is still near Paul's, since she needed to fire the rockets only very briefly, and they verified that there was only a slight difference between her clock and Paul's (Fig. 3.13). Of course, nothing would have changed unusually in Peter's clock that is far away, while she fired her rockets. Since Paul and Pauline decided, using the STR, that Peter was ageing slower and must be younger by several months by now, that conclusion doesn't change by the relatively small difference between the clocks carried by Paul

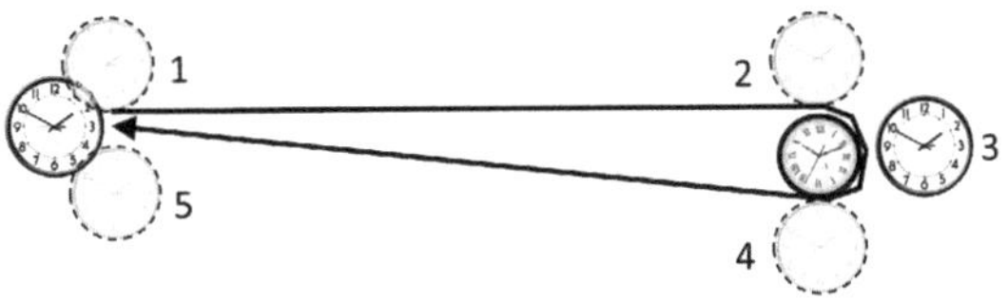

Figure 3.13 Different stages of comparing two clocks in relative motion that are initially synchronized at stage 1. One clock is non-inertial during a small part of trajectory (stages 2–4) so that the clocks can come together again for a comparison at stage 5. Since any acceleration should affect only the rate of that particular clock, such a rate-change can be tested locally, relative to another inertial clock in proximity, before and after the accelerated phase (stage 2 and 4). It is very easy to see then that none of the special relativistic 'resolutions', vouched by hundreds of textbooks, hold any water.

and Pauline. Peter, however, continues to think that the other two must be ageing slower. He also knows that there will only be a slight difference between the clocks of Pauline and Paul, if one of them fires a rocket briefly. Who really ages more and by how much? Peter or Pauline? There is still no answer in the STR.

3.11.1 Einstein's Resolution of the Twin-Clocks Problem

Realizing the need for a physical reason that can change the rate of clocks, Einstein invoked his General Theory of Relativity to break the symmetry between the twins in the STR, to solve the puzzle (Fig. 3.14). That meant his admitting that there could not be any differential ageing in the observer-symmetrical STR. If there is a differential ageing, that must be due to some other reason, to be handled by another theory.

Here is Einstein's simple gravitational argument, which he thought allowed himself to be 'extricated skillfully' from the paradox (Einstein, 1918). The time dilation in the STR is admittedly symmetric. So, Peter reckons that Pauline is younger by the factor γ, and Pauline and Paul would claim that Peter is younger by the same factor. If the relative velocity is not very high ($v < 0.1c$), we may write the time dilation during the flight duration T as $\delta T \simeq -Tv^2/2c^2 = -(L/v)v^2/2c^2$. In this expression, L is the separation after traveling for duration T at the velocity v. A comparison

DIE NATURWISSENSCHAFTEN

WOCHENSCHRIFT FÜR DIE FORTSCHRITTE DER NATURWISSENSCHAFT, DER MEDIZIN UND DER TECHNIK

HERAUSGEGEBEN VON

DR. ARNOLD BERLINER UND PROF. DR. AUGUST PÜTTER

Sechster Jahrgang. 29. November 1918. Heft 48.

Dialog über Einwände gegen die Relativitätstheorie

Von Prof. Dr. A. Einstein, Berlin.

Kritikus: Schon oft haben meinesgleichen in Zeitschriften Bedenken der verschiedensten Art gegen die Relativitätstheorie vorgebracht; nur

wegung auf den Gang einer Uhr stets Widerspruch hervorgerufen, und zwar — wie mir scheint — mit gutem Grunde. Denn dies Ergebnis scheint notwendig zu einem Widerspruch mit den Grundlagen der Theorie zu führen. Damit wir uns vollkommen verstehen, sei dies Ergebnis der Theorie zunächst hinreichend scharf angegeben.

Figure 3.14 Einstein's 'dialogue' paper on the resolution of the twin-paradox in 1918. He clarified that the symmetric Special Theory of Relativity alone could not resolve the paradox and one needed to invoke the gravitational time dilation during the phase of the acceleration, citing the equivalence between the acceleration and a gravitational field.

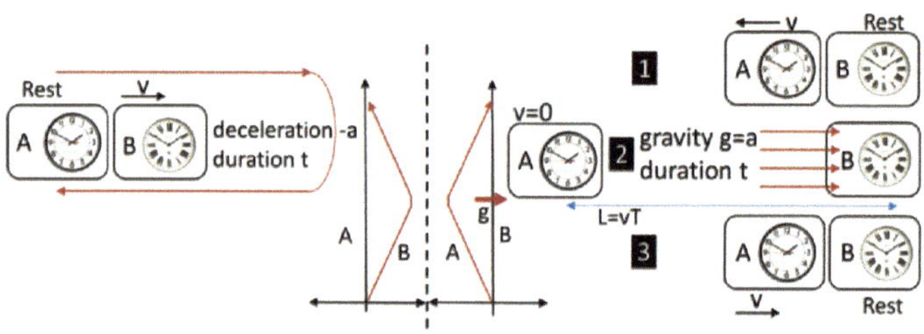

Figure 3.15 Einstein's own resolution of the twin-paradox. Relative to A, B is younger $\Delta T = -Tv^2/2c^2$. Relative to B, who decelerated to return, the total time dilation is analysed in three parts: 1 and 3) A is younger by $(T/2)(v^2/2c^2)$ in each, 2) There is brief period of gravitational time dilation, when there is a gravitational field $-g$ equivalent to the acceleration a. A ages *more* by $2Tv^2/2c^2$. Both A and B agree on the total time dilation. The space-time diagrams relative to A and B are kinematically identical. But, in B, there is a brief period when an inertial force is experienced (red arrow).

can be done only when they come together. While Pauline fires the rocket to join Peter, her deceleration is a, for a duration t, such that her relative velocity reverses, $at = 2v$. In Peter's frame, that doesn't make much difference to Pauline's ageing, if t is a relatively short duration. However, *in Pauline's rest frame, she is not moving, but just experiencing a force, indistinguishable from a uniform gravitational field* $g = -a$ *for the duration* t. Then, according to Einstein Equivalence Principle, (Fig. 3.15) and the GTR, there should be a relative gravitational time dilation, $\Delta t/t = gL/c^2 =$

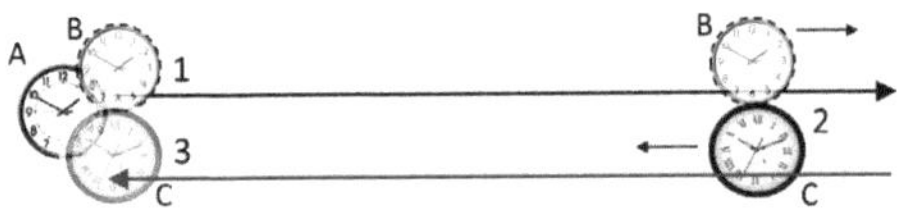

Figure 3.16 Twin-clocks comparison without acceleration! A and B are synchronized and are in relative motion. The reading of B is transferred to another inertial clock C (stage 2) that is moving in the opposite direction, which is then compared with A when C reaches A (stage 3). There is no non-inertial motion in the trajectory of either clock. Therefore, a reasoning where any part of the time dilation is attributed to the phase of acceleration, as in Einstein's resolution, is invalid.

$-aL/c^2 = -2vL/tc^2$. Note that the amount of 'gravitational' time dilation is independent of the duration of acceleration, $\Delta t = -2vL/c^2$. This is double the special relativistic time dilation δT and of opposite sign, Pauline's clock running slower than Peter's. Adding the two contributions, the total time dilation is $\delta T + \Delta t = -vL/c^2$ which is the same conclusion as Peter's; Pauline and her clock are 'younger' when she meets Peter again.

That is how Einstein solved it. Or, thought he solved it, and was confident enough to write a paper describing the solution, with the explicit statement that the claimed resolutions within the symmetric special theory were not correct (Einstein, 1918). But, he was not attentive enough, in logic. There are several reasons to point out that Einstein's resolution does not solve the paradox (Unnikrishnan, 2005b). First of all, *the twin-clocks problem can easily be posed without any accelerations*. Let A and B be two synchronized clocks that move apart at the relative velocity v. The time in B can be transferred to another clock C that is moving in the direction opposite to that of B, and finally the reading on A and C can be compared, when they are in the same location. This doesn't involve any acceleration. (see Fig. 3.16). Yet, the amount of time dilation is exactly the same as in the twin-clocks case with the acceleration. Hence, any solution that invokes acceleration is not the real solution, though it may be mathematically conforming.

Then, there is another issue that touches the real meaning of the relative time. There are two clocks in the problem and the formula is for the relative time. We just found that Pauline's clock is not affected by her brief deceleration, as she verified by comparing with the nearby Paul. Her firing rockets to alter her path cannot affect the clock with Peter, who is at a distance L away from her! So, there must be something seriously missing in Einstein's solution. Besides, take a look at some typical magnitudes within the approximation he used. If $a \simeq 10^2$ m/s for two days, and $L \simeq 10^{15)}$ m, after a year's travel at $v \simeq 0.1c$, all small compared to the numbers usually used in the textbook discussions of the paradox, then $aL > c^2$; hence Einstein's resolution does not work. Finally, there is a simple fact, almost embarrassing for any physicist who wants to invoke the 'acceleration for a short duration' cure for the paradox. Though we cannot manipulate the time of the living twins, we can always switch off or freeze the time in a physical clock during the period of acceleration! Then, the time on the clock cannot be affected by acceleration; the clock will restart with the same reading it had before the acceleration. Therefore, it is clear that acceleration has no role to play in the resolution of the twin-clocks puzzle. *Thus, we arrive at the very important conclusion that even the General Theory of Relativity does not have the correct solution to the twin-clocks problem.* The significance of this result cannot be overemphasized.

So much is written on the twin-clocks problem, most of it to dismiss the problem as a trivial misunderstanding of lesser people, and to assert that it is easily resolved within the STR. Evidently, those authors haven't noticed Einstein's 1918 paper! Ironically, the solutions discussed in such textbooks do not agree with each other – some invoke a Doppler shift of back and forth signals, some others a 'change in the line of simultaneity', and yet others invoke the mysterious role of an acceleration, without a calculation in the frame of the accelerated clock B. Our discussion has shown clearly that the problem of the twin-paradox is genuine and none of the solutions in the hundreds of textbooks and papers reaches the true resolution. There must be a completely different physical origin and reason for the time dilation of moving clocks, never discussed in physics so far. That is indeed a thrilling thought! We will soon find out that reason.

3.11.2 Time and Acceleration

There is ample experimental evidence that the relative time dilation of a clock, relative to an inertial reference clock, does not depend on its acceleration. The example that is usually given is that of the unstable particles, muons, in a storage ring accelerator; their lifetime τ is extended by a large relativistic factor γ at velocities close to c, but the large centripetal acceleration in the ring, v^2/R has no influence on the lifetime. The dilated lifetime is $\tau' = \tau/\sqrt{1-v^2/c^2}$. What is little known is the irrelevance of the large longitudinal deceleration (change of speed) in the classic example of the relativistic extension of the muon lifetime in cosmic rays. This is discussed in textbooks as a case of time dilation in inertial motion. However, the fact is very different. Due to their scattering interactions in the atmosphere, muons are moving down with the enormous deceleration of about $10^{12} - 10^{14}\ m/s^2$! Yet, the extended lifetime due to the time dilation is just $\tau' \simeq \tau/\sqrt{1-v^2/c^2}$. The firm conclusion is that the acceleration of a clock does not directly figure in the modification of time, as measured by an inertial (unaccelerated) observer.

The question that remains is about the reverse situation – time dilation of an inertial clock relative to a clock in a non-inertial rest frame. This situation is fundamentally different because the physical environment in an accelerated frame resembles the situation in a uniform gravitational field. So, how does time dilation of a moving clock differ when the reference clock is in an accelerated frame? The answer seems to be given in a straightforward way by the Einstein Equivalence Principle (EEP) and the General Theory of Relativity; the non-inertial rest frame is equivalent to a uniform gravitational field, and the relative time dilation is given by the gravitational time dilation of the GTR. This is the argument used in Einstein's 1918 paper, in his attempt to resolve the twin-paradox. Einstein asserted that it did not matter whether we were in a 'real' gravitational field or in the pseudo-gravitational field of an accelerated frame; they could not be locally distinguished by any physical means. If that was not the case, the GTR could not be completely right.

High precision experiments with modern atomic clocks, based on the microwave and optical transitions in cold atomic clouds and trapped ions, establish the gravitational time dilation in a real gravitational field, even over the small gravitational potential difference in a laboratory in the Earth's field (Chou *et al.*, 2010). Shouldn't we therefore conclude that there will be a relative time dilation between two clocks in an accelerating frame, based on the complete local equivalence of the uniform gravitational field and the accelerated frame? Nobody would expect an answer in the negative because the conclusion is based on the EEP, which is the very basis of the GTR that has been tested very well. However, the relative time dilation in an accelerated frame is not tested yet in any experiment! Such an experiment is difficult due to both conceptual and practical issues, as I will discuss soon.

Now we prove a very important and unexpected result: that *there is no relative time dilation between two spatially separated clocks in a linearly accelerated frame*. Unbelievable as it may seem, we are going to see that Einstein's Equivalence Principle is not valid for clocks in an accelerated frame. Though there is a relative time dilation in a real gravitational field, there is none in an accelerated frame.

The first step is to confirm that two clocks at relative rest in inertial motion do not have any relative time dilation. The second step is to realize that the acceleration of the two clocks for some duration T can be arranged in two ways: one by accelerating both together, or by accelerating the clocks identically, one after the other. These are equivalent since any change in the rate of a clock happens because of its own dynamical history and acceleration, without any reference to what happens to the other clocks in the world. It is only at the stage of a clock comparison, that the two clocks are brought together in proximity. *The conceptual limitation of the thought experiments designed by Einstein is in considering the relative changes as one inseparable quantity, whereas in reality all relative quantities are differences of two independently measured quantities*, with two physically separated devices, and often at physically separated locations.

There are certain results we can be sure of: (a) If the clocks are not spatially separated, they will have no relative time dilation, irrespective of the magnitude of the common acceleration. (b) If two clocks synchronized at the same location are individually, but identically, accelerated to complete a journey, one after the other, there will not be any relative dilation when they are compared at the final destination. Yet, how can a reasonable theory claim that there would be a relative time dilation if they had a finite separation at the start that is maintained at the end? How does one verify that they have a relative time dilation if they have a finite separation? Since comparison requires bringing them together, how can one be sure that the modification was due to the acceleration, and not due to the operation of bringing them together in the moving frame?!

In fact, the catch is exactly in the last point – how does one check whether there is a relative time dilation after a period of acceleration of the two clocks, if the clocks are spatially separated? One cannot. *The clocks have to be brought to the same location, for which one needs to move the clocks.* But, because of the prior acceleration, clocks are already in uniform motion, though at relative rest! Because of this, if the clocks are moved to bring them together, there would be additional motional time dilation. There is no way to separate such changes from any time dilation that might have happened during the accelerated journey, causing ambiguity in the physics of the GTR.

Even if the clocks have a common velocity before the phase of acceleration, we can deal with it. We saw that every real motion can be detected and measured accurately in the matter-filled Universe, because we have the reference of cosmic matter and radiation (Section 3.3). We have a reference system in CMBR akin to the old 'fixed distant stars', now refined and much better characterized. We can measure accurately the speed of the clocks before and after the acceleration, and calculate the motional time dilation that happens when the clocks are moved to bring them together for comparison.

Let us go ahead and calculate just the motional time dilation that happens when the clocks are brought together for comparison, after *explicitly assuming that there is no relative time dilation of spatially separated clocks in an accelerated frame.* If our calculation of just the motional time dilation, which happens much after the acceleration is completed, agrees completely with the GTR prediction of the pseudo-gravitational time dilation due to the acceleration, then we can be sure that the time dilation has nothing to do with the EEP and the equivalent gravitational field. This would imply that the GTR assertion of the gravitational time dilation in accelerated frames is not correct. We will see at once that the GTR incorrectly ascribes the motional time dilation during the act of comparison to the relative time dilation due to a pseudo-gravitational field during the phase of acceleration!

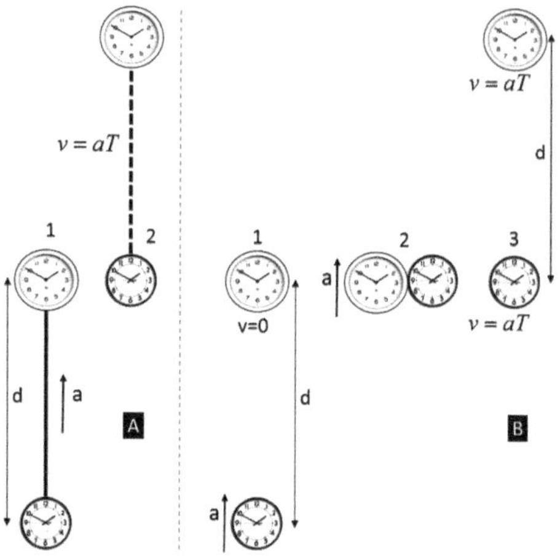

Figure 3.17 Is there a relative time dilation for two identically accelerated clocks, when they are spatially separated? (A) Two clocks separated by a distance d are accelerated up together for the duration T. (B) The same experiment done in two steps: first the lower clock is accelerated for the duration T. When the two clocks are together, they may be compared to verify the motional time dilation (at stage 2). Then the upper clock is identically accelerated. The initial and final configurations are the same in both (A and B) experiments, with the two clocks at relative rest. To compare the reading of the two clocks, they need to be brought together, in both experiments.

Let us start with two clocks C1 and C2, separated by the distance d in a laboratory, stationary relative to the reference cosmic frame (Fig. 3.17A). Since we are sure that both clocks are at zero (absolute) velocity, there is no difficulty synchronizing them independent of any theory or convention. Then the laboratory moves at the acceleration a for some duration T (say, as measured in one of the clocks). Both clocks will have the time dilation, but they are identical, since we assumed that there is no additional relative time dilation in an accelerated frame (denying the GTR assertion). After the phase of acceleration, the laboratory and the clocks are moving at the speed $v = aT$; this may be verified by measuring their speed relative to the cosmic frame. To compare the clocks, at least one of the clocks needs to be moved through the distance d, to bring them together. Let us move the lower clock C1 slowly at the small velocity v_c. The absolute velocity of C1, relative to the cosmic frame, is then $v + v_c$. The two clocks will be together for the comparison after moving for the duration td/v_c. The time dilation of the transported clock due to the motion is

$$\Delta t_A/t_A = -(v+v_c)^2/2c^2 = -(v^2+v_c^2+2vv_c)/2c^2 \qquad (3.22)$$

The time dilation of clock C2 that is moving with the laboratory at the velocity v, in the same duration, is

$$\Delta t_B/t_B = -v^2/2c^2 \qquad (3.23)$$

Since we have chosen v_c very small, $v_c \ll v$, the time dilation of clock C1, relative to clock C2, is

$$\frac{\Delta t_A}{t_A} - \frac{\Delta t_B}{t_B} \simeq -vv_c/c^2 = -aTv_c/c^2 \qquad (3.24)$$

Since the duration of slow transport $t = t_A = t_B$ we write this as

$$\frac{\Delta t_A - \Delta t_B}{T} = -atv_c/c^2 = -ah/c^2 \qquad (3.25)$$

This is the motional time dilation that happens in the act of bringing together the clocks for the comparison. It has nothing to do with what may have happened during the acceleration earlier. However, this expression matches exactly with the expression for the gravitational time dilation of the two clocks in the GTR, whereas we have explicitly assumed here that there is no such time dilation in the accelerated laboratory. This relative time dilation, calculated in a straightforward manner as the inevitable motional time dilation, is what the GTR incorrectly interprets as the time dilation in an accelerated frame, due to the assumed and postulated equivalence between an accelerated frame and a gravitational field. Since we have the cosmic references to verify whether we really moved, there is no confusing between the accelerated motion and a gravitational field. *We have just proved that Einstein's Equivalence Principle is totally violated in the case of the time dilation of clocks in an accelerated frame. There is no relative time dilation for spatially separated clocks in an accelerated frame.* If a real experiment is done, all that will be measured is the motional time dilation after the acceleration, entirely due to the additional motion required to bring the clocks together for the comparison.

The experiment can be done in two stages, where each clock is independently accelerated. First the lower clock is moved at the acceleration a, for the duration T. When the clocks are together, they may be compared to verify that the moved clock has just the motional time dilation. Then the second clock is accelerated identically. The initial and final configurations, as well as the relative time dilation, in the experiments A and B of Fig. 3.17 are exactly the same.

To summarize the situation, there is the gravitational redshift of light that is propagating in a gravitational field, and there is the Doppler redshift of light in an accelerated laboratory. There is a relative time dilation of clocks in a gravitational field. But, there is no relative time dilation of clocks in an accelerated laboratory.

We have seen that neither the Special Theory of Relativity nor the General Theory of Relativity offers the correct framework to deal with the empirically observed fact that clocks have their rate modified in inertial as well as non-inertial motion, and that the modification does not depend on the amount of acceleration. But, these are the only theories we have at present to deal with the physics of clocks and motion! We have come face to face with the situation that our best theories, matured over a century, are not enough to understand an everyday phenomenon like the motional time dilation of a clock, nowadays a routine fact in applications like the GPS. That is a difficult situation for a physicist, as well as for the society. Well, I would say it is an exciting situation, because clearly we are in the vital need of a different and new theory!

3.12 Time in Rotating Frames

It is often said, correctly, that the time dilation of clocks depends only on their speed and that acceleration is not important for determining the magnitude of the time dilation. But, acceleration is the thin thread people hold on to in debates on time dilation, while deciding which clock of the two in relative motion ends up recording less time. Generally people are careless to mention the frame in which the statements about time dilation and acceleration are made.

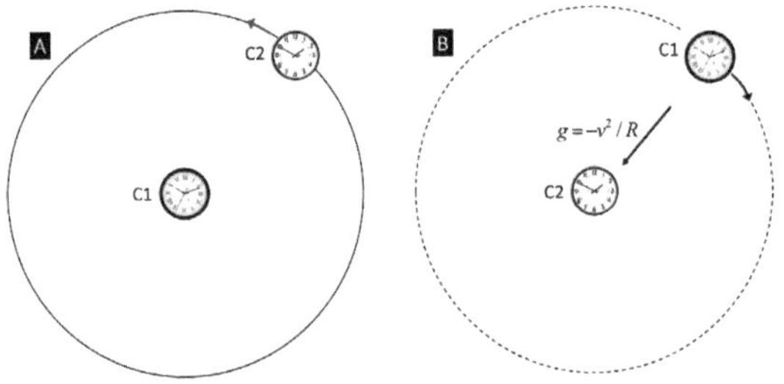

Figure 3.18 Time dilation of a clock in non-inertial motion, from two frames. (A) Inertial frame of the reference clock. (B) The rest (proper) frame of the accelerated clock.

Let us look at one example. One situation is when a clock C2 is taken around at uniform speed v in a circle of radius R, as in Fig. 3.18. There is no gravity. We have seen that the time dilation of such a clock, relative to another one (C1) at rest at the centre, $r = 0$, is $\delta T \simeq -Tv^2/2c^2$, in the STR as well as in the LPTR. Such time dilations are routinely verified in experiments (in a storage ring for unstable elementary particles, for example). The centripetal acceleration of the clock might be large, but that is not relevant for the time dilation. Or, is it?!

The answer given in the STR and the GTR depends on the frame in which the statement is made. If we are in the same frame as the frame of the moving clock C2, there is no relative motion. Instead, there is a radial force that is indistinguishable from a strange gravitational force, $g = -v^2/R = -R\omega^2$. 'Strange' because it is radially inwards with an apparent source at the location of the clock C2 (Fig. 3.18B). Then, the *formal gravitational time dilation* of C2 relative to C1 is $\delta T_g/T = gR/c^2 = -v^2/c^2$. The clock C2 goes slower by this amount relative to the clock C1. However, the clock C1 itself is now in motion at the speed v, relative to the rest frame of C2! So, it must be going slower, according to the STR, at the rate $\delta T_v/T = -v^2/2c^2$. Though the clock C1 is in the same (pseudo) gravitational field, it does not have any gravitational time dilation because it is 'freely falling'(!). Therefore, the total relative time dilation is $\delta T/T = \delta T_g/T - \delta T_v/T = -v^2/2c^2$, the clock C2 going slower. The result is the same (as it should be) in both frames, but the *physical reasons*

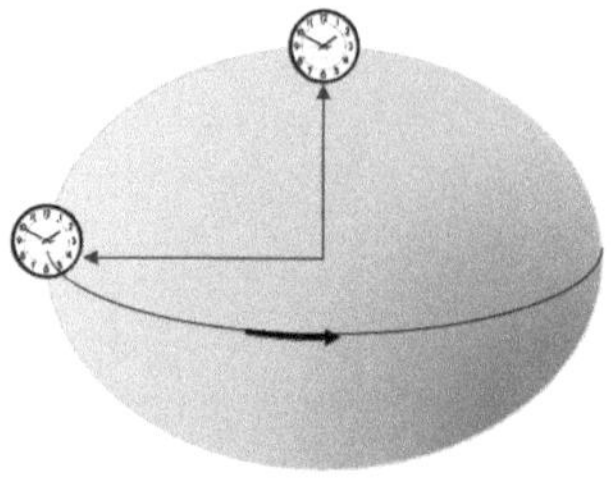

Figure 3.19 Time dilation of clocks on the rotating Earth. The motional time dilation is maximum at the equator, and zero at the poles. But the gravitational time dilation is larger at the poles.

for the time dilation are different. In one frame it is just the relative motion, and in the other it is the relative motion and gravity.

3.12.1 Clocks on the Geoid of the Rotating Earth

A clock that is stationary in a laboratory on the Earth's surface will have the motional time dilation as well as the gravitational time dilation. The motional time dilation due to the constant speed of rotation $v = R\omega$ is $\delta T/T = -R^2\omega^2/2c^2$. This is a maximum at the equator and decreases to zero at the poles. The gravitational time dilation is expected to be $\delta T_g/T = -GM/Rc^2$ relative to the situation with no gravity. How does the total time dilation of a clock at the equator compare with that of a clock at the poles? It turns out that the gravitational time dilation of a clock at the poles is larger than that of a clock at the equator, due to the centrifugal bulging of the Earth (Fig. 3.19). In fact, the shape of the Earth was determined by the hydrodynamic balance of gravity and rotation, when it was in a more fluid state, from which it was solidified. Therefore, the sum of the gravitational potential and the (pseudo) centrifugal potential is the same everywhere on the surface of the Earth (but for the variations due to the density variations, local undulations as mountains, etc.).

The centrifugal force due to the rotation of the Earth is given by $F_c = v^2/r = r\omega^2$. This can be written as the gradient of a pseudo-potential by writing $F_c = -dV_c/dr$. Then, $V_c = -r^2\omega^2/2$. Then the quantity $\phi_g + V_c = GM/r + r^2\omega^2/2$ is the same everywhere on the mean surface of the Earth. Clearly, the motional time dilation can also be considered as a pseudo-gravitational time dilation in the rest frame of the Earth by writing it as $\delta T_c/T = V_c/c^2 = -v^2/2c^2$. Therefore, the total time dilation, $\delta T/T = \phi_g/c^2 - v^2/2c^2$ is the same everywhere on the mean surface of the Earth. The clock at the equator, which moves at about 460 m/s due to the Earth's rotation, does not go slower than a clock at the pole, because the latitude dependence of the gravitational time dilation compensates the latitude dependence of the motional time dilation. (Six years before discovering the gravitational time dilation, Einstein wrote in the 1905 paper on the STR,

> Thence we conclude that a balance-clock at the equator must go more slowly, by a very small amount, than a precisely similar clock situated at one of the poles under otherwise identical conditions.

This expectation was negated by his own discovery of the gravitational time dilation.)

3.13 Motion, Gravity, and Time

We have seen that the physical time measured by clocks is affected in two distinct ways. One is through motion, and the other is by the gravitational interaction. While the gravitational time dilation posed no conceptual issues, our discussion of the motional time dilation raised more questions than answers, leaving us in a thoroughly unsatisfactory situation. We could not identify any physical cause for the motional time dilation of clocks in relative motion. We saw reliable proofs of the incorrect nature of the standard resolutions of the twin-clocks problem. We also saw in detail how Einstein himself rejected these standard textbook resolutions within the STR. Then we saw how Einstein's solution that invoked a pseudo-gravitational field during the acceleration fell apart easily, in multiple ways. In short, the real physical cause of the ubiquitous motional time dilation must be some new fact of physics, not recognized as yet. What is that elusive physical cause?

Perhaps there are not two *different* causes – motion as well as gravity – for the phenomenon of time dilation! After all, the modification of time is a physical change in the clocks, and perhaps only the gravitational interaction can induce such a universal modification. Motion might be just an indirect agent that brings about a difference in the gravitational interaction. That sounds very reasonable. But, gravitational interaction with what? Such easily observable time dilation requires the gravitational interaction with an enormous amount of matter.

The answer was not obvious in the early 1900s. Even a speculative guess was not attempted either, except the remarkable conjecture by Ernst Mach that the elusive origin of the inertial forces might be in some interaction with the matter in the Universe. Not much matter was seen then, to take such suggestions seriously. But now, there is no excuse.

3.14 It Must Be Gravity's Time

Sometimes the truth is so transparent that we fail to see it even when repeatedly and persistently pointed to. We are well aware of the fact that the proximity to matter and its gravitational potential affects the rate of clocks. We know that there is an enormous amount of matter in the Universe, sourcing the dominant gravitational potential at every place in this Universe. We know that any interaction potential assumes a velocity dependent value in a frame moving relative to the sources of the potential. Yet, we fail to connect the thick and closely spaced dots to deduce that the matter in the Universe and its enormous gravitational potentials are the reason for the velocity dependent rate of clocks that move relative to the cosmic matter! The only 'reason' for this unreasonable stance is that the moment we make this straightforward deduction, the entire prevailing edifice of the physics of relativity and dynamics we believe in, the Special Theory of Relativity, will cease to exist and would be replaced by new physics as well as new philosophy. Faith and confidence built up over a century will collapse entirely. Is it some fate that needs to be dreaded or encouraged? Is there anybody who can deny this plain truth with any of the aids that science provides – logic, observations, calculations, etc. The answer is a resounding NO; yet, a century has passed with many compelling clues parading in front of our eyes repeatedly.

This century is the century of cosmology. We are aware of the outstanding problems that we need to tackle to claim a mature understanding of our Universe. But, one thing we are absolutely sure of is the presence of a significant amount of matter and energy – sources of gravity – in this Universe. And such a mighty source as the entire matter in the Universe will have an enormous gravitational influence on all physical processes. Given this situation, our clocks cannot be veiled from the gravity of this matter and energy. All clocks, and hence time itself, must be significantly affected by cosmic gravity. It must be Gravity's Time.

3.15 Summary

In this chapter, I discussed the relation between motion and the modification of time measured by clocks. This needed clarifications on the distinction between relative and absolute motion as well as between inertial and non-inertial motion. I mentioned the Galilean Principle of Relativity (PoR) and its generalization to the phenomena in electrodynamics, including the propagation of light. The failure to detect the hypothetical stationary medium of the ether, in several experiments including the Michelson–Morley (M-M) experiment, led to the theory of relativity by Lorentz and Poincaré, with new physical inputs like the length contraction and time dilation that were intertwined with the Galilean propagation of light. The mathematical description of the theory was based on the Lorentz Transformations (LT). Einstein's Special Theory of Relativity, with the same PoR and a radically different new hypothesis of the invariance of the relative velocity of light in all inertial frames, changed the course of physics and the theories of dynamics. I emphasized the paradigm shift to a theory in which all physical effects were linked to the relative motion between two frames, with no more any meaning to the notion of absolute rest or real motion.

Next, we examined in great detail the propagation of light relative to inertial frames, to compare and contrast the Galilean propagation with a variable relative velocity and the Einsteinean propagation with an invariant relative velocity. A thorough study of the M-M experiment and its variations showed that the null result was consistent with both, because only second-order (v^2/c^2) effects are expected in all experiments with the two-way propagation of light.

Then, I pointed out the factual existence of an absolute reference for motion in the matter-filled Universe and the Cosmic Microwave Background Radiation (CMBR). Besides serving as a reference for absolute motion and velocity, the slowly decreasing average temperature of the CMBR serves as a universal clock that is globally synchronized.

This brought us to an extensive and complete discussion of the notion of simultaneity in space and time, comparison of clocks, and the synchronization of spatially separated clocks. The methods of clock synchronization by electromagnetic signals as well as by the physical transport of the clocks were studied in detail, and the relevant mathematical expressions were derived and discussed. These were then compared to the modifications indicated in the LT, to bring out clearly the intertwined nature of the motional modification of time and the assumptions on the propagation of light. I proved that the two cannot be disentangled in any two-way experiment or in any one-way propagation experiment that uses spatially separated clocks for the measurement.

After briefly reviewing the representation of events in space and time with the help of a space-time diagram, we embarked on a unique and complete discussion of the much-debated twin-paradox, or the asymmetry in the twin-clocks problem, revealing many aspects that were brought together for the first time in a logically and physically consistent discussion. In particular, the standard 'resolutions' within the symmetric theory of Special Relativity were compared to Einstein's own conflicting resolution that used the postulated equivalence of an acceleration and a gravitational field in his General Theory of Relativity. The discussion exposed the easily verifiable inconsistencies and showed the pressing need to understand the *real physical reason behind the phenomenon of motional time dilation*. It became clear to us that no resolution based on the physics in an empty space, devoid of gravity, offers a consistent and general solution to the twin-clocks problem. This was in immediate contrast to the factual situation that the real arena of physical processes is the enormous Universe filled with matter and energy, and not the empty space assumed in the prevailing theories of relativity.

This century is the century of observational cosmology. If we acknowledge the observed fact that all physical processes and their measures happen in the perennial presence of the gravity of the matter and energy in this Universe in which we live, then the obvious and only consistent physical conclusion is that all the clocks, and hence time itself, must be significantly affected by cosmic gravity. This conclusion and its quantitative consequence are unavoidable within the physical principles and empirical results we already know. It must be Gravity's Time.

Chapter 4

Our Universe and Cosmic Gravity

Our view of the Universe evolved very slowly, taking thousands of years to mature. While tremendous advances were made in the scientific understanding of the physical world in our neighbourhood, the larger Universe remained inaccessible for several millennia. It is in some sense unbelievable that the knowledge we held about the contents and extent of the Universe outside the solar system did not change much for about 2000 years, from the Greek focus on astronomy and mathematics to the beginning of the 20th century. During this period, we learned much about the solar system, its planetary motions, and our place in it. Physical science and the description of dynamics were seeded and developed to its present mature form.

Yet, our impression and understanding of the larger Universe remained essentially the same for many centuries, till about 1930! For us, the vast Universe outside the solar system consisted of the innumerable distant stars spread over the whole sky, the Milky Way, which was identified as our galaxy of billions of stars, and a few nebulae sprinkled in the deep celestial sphere; this completed our Universe till then. Even the invention and the development of telescopes did not change the situation significantly, as far as the knowledge of the extra-galactic Universe was concerned. The

Gravity's Time
C. S. Unnikrishnan

ISBN 978-981-4877-72-5 (Hardcover), 978-1-003-25656-4 (eBook)
www.jennystanford.com

scenario changed drastically with the installations of the new big telescopes in the early 20th century, aided by the improved techniques of photography and the advances in recording and interpreting the atomic spectra.

A new and very different Universe emerged as a revelation after Edwin Hubble's observations with the newly commissioned 100 inch diameter telescope at Mount Wilson in the USA. Hubble published his findings and their interpretation in 1929 (Hubble, 1929). We became aware of the vast distribution of extra-galactic matter, in numerous other galaxies that are receding from us with their velocity approximately proportional to their distance. The matter-filled Universe not only was very vast but also was dynamic and expanding steadily.

This was a timely discovery, when there already were several theoretical attempts at describing a Universe with matter distributed throughout the space, applying Einstein's General Theory of Relativity. While Einstein himself wrote about a static Universe, obtained by modifying his original equations, there were others who discussed a dynamical Universe that was expanding or contracting. Hubble's observations determined the future course of cosmology (Ryden, 2002).

But, the findings in observational cosmology did not touch the course of physics, except in some of its speculative spheres. This was unfortunate, because modern cosmology probed and established the matter content of the vast Universe through direct observations as well as by studying the gravitational effects of that matter. *Yet, we never considered the obvious and large gravitational effects of the cosmic matter in the physics of relativity and dynamics.* This situation cannot be defended logically, or empirically.

We are now ready to examine the relation between motion, gravity, and matter in the Universe, and how all these affect the passage of time, the rate of clocks. Then, we will realize that dynamics, relativity, and the gravity of the matter in the Universe, all are parts of a single coherent paradigm that holds the entire edifice of physics, as a single whole. *I have named this gravitational paradigm of physics and the theory of dynamics and relativity based on cosmic gravity as Cosmic Relativity* (Unnikrishnan, 2009). This theory of relativity is the only one that acknowledges and incorporates the gravity of the matter in the Universe in its structure; therefore, it is the only theory that is consistent with the features of the factual Universe, in which all physical processes and the experimental tests of theories happen. The theory agrees with all empirical facts to date, and it is also fully supported by new experimental tests. The topic of time and its modifications is but a facet of this coherent edifice.

4.1 Leaning on Galileo

Being in Pisa is a moving experience. The cathedral holds an idea of a pendulum clock in its chandelier, near the leaning tower that symbolizes the Universality of Free Fall as well as gravity's mundane action that can bring even grand structures down.

Galileo discovered, and asserted with elaborate reasoning, two fundamental principles that evolved to become cornerstones of physics, and remained so. One is the universality of the rate of fall or acceleration under gravity, which became the Principle of the Equivalence of Inertia and Mass, now called the Weak Equivalence Principle (WEP, to distinguish it from the more general EEP). Another is the Principle of Relativity – the impossibility to detect uniform (inertial) motion at any speed by local experiments. The deep fact that these two principles are intimately related is not yet realized by the current paradigm of physics, which does not acknowledge cosmic gravity as the determining factor of all dynamics and relativity. These connections will become clear as we progress logically on this exploration.

We have already discussed the Principle of Relativity (PoR), which is part of our daily experience. While in a smooth uniform motion, we cannot feel or measure any motion. All physical processes, and hence all laws of physics, remain exactly the same as in a frame at rest. After all, we have no physical sensation of any of the motions of the Earth, around the Sun or in the galaxy along with the Sun, even though we know from other evidence that the speed of the motion is a stupendous 400 km/s or so. This 'invariance' property of physics in inertial motion is the statement of the PoR. Poincaré and Einstein generalized it to include electrodynamics, and indeed all of physics.

The mathematical statement of the Galilean PoR is that physical processes are invariant under the Galilean Transformations of the coordinates,

$$x' = x - vt \tag{4.1}$$

$$t' = t \tag{4.2}$$

Clearly, Newton's law of motion, $\frac{d^2\vec{x}}{dt^2} = \vec{a} = F/m$, is invariant (retains the same mathematical form) under these transformations, when the velocity is constant. Then, changing the frame of reference from x, t to another inertial frame x', t' at the relative velocity v does not contribute to the acceleration $a = F/m$.

What is the *physical reason* behind the PoR? *There is no answer in current physics.* That is why it is stated as a hypothesis, or as a 'principle'. However, we will see a precise answer within my gravitational paradigm of Cosmic Relativity that defines the correct and consistent physics of relativity.

This brings us to the next principle – the Universality of Free Fall (UFF). It is the statement that all bodies fall at the same rate in a gravitational field, irrespective of any material properties. In particular, all bodies fall with the same acceleration independent of their mass. When Galileo realized this, he studied the amazing fact in great detail, even with several experiments. It is indeed a profound discovery, as realized later by Newton.

Newton discovered and formalized both the precise law of dynamics as well as the inverse square law of the gravitational interaction. These are two independent findings, connected only by the common discoverer and a mystery. The mystery is that seemingly the same physical measure of matter appears in both the laws – the mass. In the law of dynamical acceleration, applicable to all kinds of forces and interactions, it appears as the 'inertia' to motion, $\vec{a} = \vec{F}/m_i$. In the law of gravitation, it appears as the source (charge) of gravity, $F_g = -GMm/R^2$, in the expression for the force between two bodies of (gravitational) mass M and m.

Therefore, when we consider the dynamics of free fall in a gravitational field, applying the law of dynamics, we get the gravitational acceleration as

$$a = F/m_i = -\frac{GMm}{R^2 m_i} \tag{4.3}$$

If the mass of the falling body m is the same physical quantity as its inertia m_i, then $a = -GM/R^2$, which is independent of the mass m. That the free fall is independent of other material properties is merely a reflection of the fact that both dynamics and the gravitational force do not care about any other material property. Thus, Newton found that the explanation of Galileo's UFF is in identifying mass and inertia of a body. It may be recalled that the equivalence of the gravitational mass and the inertial mass is the Weak Equivalence Principle. These topics were discussed briefly in Chapter 2 (Section 2.3).

What is the *physical reason* for the equality of mass and inertia? There is no answer in today's standard physics. But the cosmic gravitational paradigm, and the theory of Cosmic Relativity, will provide a clear answer.

4.2 Matter, Gravity, and Metric

Modern theoretical cosmology is entirely based on the General Theory of Relativity (GTR), which is Einstein's theory of gravity. That is, a model of the Universe that is consistent with the observational features of our Universe is built based on a solution of Einstein's equations (EE) of the GTR. These equations are very general and each of their solutions - expressions that describe spatial distances and temporal intervals in a gravitational field, or the 'metric', as a function of coordinates - corresponds to some physical situation. The difference between the square of the spatial distance $dx^2 + dy^2 + dz^2$ and the squared time interval dt^2 (scaled with the constant c^2) is defined as the square of the 'line element'. We have seen this quantity earlier, when we discussed the relativity of inertial motion. In a general form, the line element is the quadratic expression

$$ds^2 = \sum_{i,j} g_{ij} dx^i dx^j \tag{4.4}$$

where the 16 coefficients g_{ij} constitute the metric (here, 'x' stands for any coordinate, and the superscript i, $j = 0, 1, 2, 3$ distinguishes t and the three spatial coordinates). The constant c is included with the coordinate t, such that the units of both time and space are the same (that of the length). Since the metric specifies how the spatial and temporal intervals behave in the presence of matter, these quantities represent the gravitational potentials in the GTR. For example, in the familiar situation of a location near the Earth, at distance r from the centre, $g_{00} = -(1 - 2GM/c^2r)$. Therefore, $\sqrt{-g_{00}} - 1 = \sqrt{1 - 2GM/c^2r} - 1 \simeq -GM/r$ is the Newtonian gravitational potential.

There are many components of the 'metric tensor' g_{ij}, 16 in all. But, changing the order of the coordinates in the quadratic expression does not change the physical situation in our familiar space and time; $dxdy$ is the same as $dydx$, and $dxdt$ is the same as $dtdx$. So, the tensor is 'symmetric', $g_{ij} = g_{ji}$. Now, there are 10 *independent* components in all, and there are as many gravitational potentials in general. However, the proliferation is mostly due to the number of dimensions, the same kind of potential repeating for each dimension. We can focus on the nature of these potentials by limiting our attention to one space dimension, say $x^1 \equiv x$, and time ($x^0 \equiv t$). Then, there are only three independent metric components, corresponding to the coordinate products c^2dt^2, dx^2, and $cdxdt$. These are g_{00}, g_{11}, and $g_{01} = g_{10}$. The first and the second are related to the familiar Newtonian potential, $-GM/r$, of a static mass source. The third, that has both space and time coordinates, involves motion and its velocity, because that is the only way to have both space and time mixed.

Since this is an important point that is rarely explained, let me indicate how this comes about, considering only one spatial dimension. The spatial line element at rest is $dl^2 = dx^2$. Now, I consider the Galilean motion $x' = x - vt$. Then $dl'^2 = dx'^2 = dx^2 + v^2dt^2 - 2vdxdt$. So, though we started with a metric with only one purely spatial component g_{11}, mere uniform motion has resulted in a line element with a term (v^2dt^2) that modifies g_{00} and a mixed term with the velocity $-2vdxdt$, which corresponds to g_{01}. We will soon see the dramatic physical consequences of this simple fact.

We see that the gravitational metric is not difficult to understand. Just as in electromagnetism, *there are only two fundamental entities as sources for the fields and physical effects: the charge and the current* (of the charge). These are the only material realities. The motion of the charges is the current. The electric charge generates an electric potential ϕ and an electric field $\vec{E}$, and the current a 'vector potential' $\vec{A}$ and a 'magnetic' field $\vec{B} = \nabla \times \vec{A}$. There are no more fundamental categories; the charges and the currents exhaust the possibilities of the sources of fields.

In gravity, 'mass' is the charge that generates the 'electric' type familiar gravitational field – the pulling force. The current of masses then should generate a gravitational vector potential and a 'magnetic' type gravitational field, which is a new type of field in relativistic gravity. We shall call it, following the standard convention, a gravitomagnetic field, or gm-field. Then, the metric components g_{0i} are gravitomagnetic potentials, proportional to the velocity of motion. Therefore, whether one is dealing with electromagnetism or gravity, the sources for all physical phenomena are only of two kinds – charges and their currents. Gravity is more complicated because of the equivalence of energy and mass, which makes all forms of energy also the charge of gravity.

The presence and distribution of matter and energy determine the metric. For example, the empty space has a description that obeys Einstein's equations: it is expressed as $ds^2 = dx^2 + dy^2 + dz^2 - c^2dt^2$, which we have already seen as the Minkowski line element. Choosing a convention, we write the time coordinate as the 'zeroth' coordinate and the three spatial coordinates are labelled 1 to 3. Then the Minkowski metric is diagonal, $g_{00} = -c^2, g_{11} = g_{22} = g_{33} = 1$ and the rest of the components are zero. In terms of its symmetries, this metric is *homogeneous and isotropic*, apart from having zero curvature of space and time, as one would expect for space with no matter. It represents space that is the same everywhere (homogeneity) and the same in every direction (isotropy). The metric can also be written as a matrix array for the four coordinates of space and time (with t-coordinate first, conventionally) as

$$\begin{pmatrix} -c^2 & 0 & 0 & 0 \\ 0 & 1 & 0 & 0 \\ 0 & 0 & 1 & 0 \\ 0 & 0 & 0 & 1 \end{pmatrix} \qquad (4.5)$$

Since empty space remains the same everywhere and always, the metric has no dependence on the location or the epoch of time –

it does not depend on the coordinates. If we move in empty space, nothing changes, and the motion cannot even be detected, because there are no references for motion. *Empty space remains homogeneous and isotropic even for a moving observer, and this is the entire basis of the STR and its Lorentz Transformations.* As I repeatedly emphasize, such a world does not exist, even as an approximation to reality.

The Universe with its matter distributed uniformly and without any coherent motion is another solution of the Einstein's equations, with the properties of being homogeneous and isotropic, albeit only in non-moving frames. However, when there is matter, the dependence of the metric on the coordinates reflects that fact (Ryden, 2002). The line element for the Universe, discovered and interpreted in the works of A. Friedmann, G. Lemaître, H. P. Robertson and A. G. Walker (FLRW line element), is

$$ds^2 = -c^2dt^2 + a^2(t)\left(\frac{1}{1-kr^2}dr^2 + r^2d\theta^2 + r^2sin^2\theta d\phi^2\right) \quad (4.6)$$

This can be compacted as $ds^2 = -c^2dt^2 + a^2(t)d\Sigma^2$, where Σ represents the spatial part of the line element. $a(t)$ is the 'scale factor' that represents the evolution of the spatial scale of the Universe. The line element simplifies considerably when the 'spatial curvature' represented by the constant k is zero, as it is verified from modern observations for our Universe, which is the only known Universe. (The curvature term 'k' is related to the density of matter and energy in the Universe. Therefore, $k = 0$ implies a particular evolution of density, called the critical density, which is indeed the observed evolution.) Then the line element written in terms of the more familiar Cartesian coordinates is

$$ds^2 = -c^2dt^2 + a^2(t)(dx^2 + dy^2 + dz^2) \quad (4.7)$$

with the homogeneous and isotropic time dependent metric

$$\begin{pmatrix} -c^2 & 0 & 0 & 0 \\ 0 & a^2(t) & 0 & 0 \\ 0 & 0 & a^2(t) & 0 \\ 0 & 0 & 0 & a^2(t) \end{pmatrix} \quad (4.8)$$

When the rate of change of the scale factor ($\dot{a} = da/dt$) is positive, it represents an expanding Universe, in which the spatial distances increase with time. This model fitted very well Edwin Hubble's early observations of the steady recession of all distant galaxies, with their velocity nearly proportional to their distance, $v = Ht$. Here, $H = \dot{a}/a$ is the Hubble constant, with its present measured value nearly 70 km/s per Megaparsec (Mpc). One Mpc of distance is 3.1×10^{22} m. That is, if the distance to a galaxy is a megaparsec, it should be moving away from us at the speed of about 70 km/s. That translates to just a tiny change of about a micron/second on the scale of the solar system. Thus, the rate of expansion is very slow at present. If we choose the units such that the value of $a(t)$ at present is taken as 1, then the scale factor can be written as $a(t) \simeq 1 + Ht \simeq 1 + 2.3 \times 10^{-18}t$, where I have written the value of the Hubble constant in the 'laboratory' units 'm/s per metre'. Therefore, it is clear that laboratory-scale experiments will be insensitive to the time dependent features of the metric and the line element, even with extended observations. Hence, on the laboratory and terrestrial scales, $a(t)$ can be considered as a constant.

Thus, the smallness of the Hubble parameter results in the *superficial similarity* between the Minkowski metric of empty space used in Einstein's relativity and the FLRW metric of the factual $k \simeq 0$ Universe, in frames at rest relative to the average matter distribution.

$$\begin{pmatrix} -c^2 & 0 & 0 & 0 \\ 0 & a^2(t) & 0 & 0 \\ 0 & 0 & a^2(t) & 0 \\ 0 & 0 & 0 & a^2(t) \end{pmatrix} \simeq \begin{pmatrix} -c^2 & 0 & 0 & 0 \\ 0 & 1 & 0 & 0 \\ 0 & 0 & 1 & 0 \\ 0 & 0 & 0 & 1 \end{pmatrix} \tag{4.9}$$

This will turn out to be very important in understanding why we have been misled for a century about dynamics and relativity, by a theory built on the empty space without matter, while factually living in the enormous gravity of all the matter in the Universe.

There are several other interesting solutions of Einstein's equations. What is important to note is that each is applied in a particular situation. The FLRW solution is used while dealing with dynamics and the modelling of the Universe, but not while dealing with moving clocks in the laboratory; there we use the Minkowski empty space solution. Thus, in the latter situation, we forget the fact that we are moving the clocks in the very same Universe with matter! We failed to recognize that the Universe is a very special and unique physical situation that encompasses everything, including all our experiments and our life. Einstein's equations are operative within this Universe, and so are all experimental tests. There is no empty space, and the Minkowski metric is an impossible and non-existent situation, even as an idealized approximation. The superficial similarity between the two metrics mentioned above is thoroughly misleading, because the similarity is valid only for an observer who is at rest relative to the matter distribution in the Universe! We will see, and be enlightened, that the metric, and therefore the gravitational potentials are very different from those represented by the Minkowski metric, for an observer or a clock in motion in the factual Universe. That is the crux of the matter.

4.3 Contents of Our Universe

After the paradigm-changing interpretation of Hubble's discovery, the next 50 years of observations with the large telescopes established that the matter in the Universe was distributed more or less uniformly and isotropically everywhere. The recession of the galaxies implied a dense hot past, and a possible origin termed as the 'Big Bang'. The minimum age of the Universe was estimated as 14 billion years, from the age of the oldest structures visible in the Universe and the model details of the Big Bang Universe. It was realized that there was much more matter than what was observable optically as well as with the radiation in other parts of the electromagnetic spectrum. This was deduced from the gravitational dynamics of the stars and gas in the galaxies. In fact, it was realized that most of the matter in the Universe does not emit light; the dark matter dominates the matter distribution in the Universe. The theoretical picture became clearer with much more accurate measurements of the rate of recession of the distant galaxies. An accurate estimate (within 3%) of the average density of the matter in the Universe became possible. The presently estimated value of the density is about 10^{-29} g/cm^3, which is the value consistent with a vanishing spatial curvature, at the critical density; the 'k' parameter in the FLRW line element is observationally consistent with being zero.

There are still many puzzling features about the matter in the Universe (Ryden, 2002). Less than 5% of all matter is the familiar matter consisting of protons, neutrons, electrons, light, etc. (we must include energy as matter, remembering that both are equivalent). At least 30% of the cosmic matter is the invisible 'dark matter' that is clustered around the galaxies and groups of galaxies. We do not know yet its composition. However, the estimate is reliable, having measured through its gravitational pull on the stars and gas, as well as from its gravitational bending effect on light. The remaining 65% of the invisible cosmic matter, estimated from the measurements on the recession of the galaxies in the Universe at very large distances, is a very deep mystery. It is tentatively called the 'dark energy' and has a long theoretical history. Einstein had introduced the notion in the early years of his GTR, as a geometric entity (and not as matter) called the cosmological constant, in his equations to 'construct' a static Universe. Hubble's observations forced the rejection of the idea. Apparently, Einstein had even regretted about his 'greatest blunder'. However, modern observations seem to suggest the necessity of a smoothly distributed, gravitationally active invisible matter or energy, with rather strange physical properties resembling Einstein's cosmological constant.

All this points to one very significant fact, that the entire Universe as well as our neighbourhood is bathed in the enormous gravitational potential due to the matter and energy in the Universe. But, this all-important physical fact is what remains missed, and continued to be ignored, even after being pointed out! Unbelievably, nobody acknowledges this crucial fact and its obvious importance for dynamics and relativity. Perhaps the reason for this is that admitting this realization that came after we 'completed our theories' would demand an immediate and radical revision of the 'established' theories. That would be a change as drastic in comparison as the shift from the geocentric view to the heliocentric view of the solar system. But, *one cannot deny an already observed Universe, its vast amounts of matter, and its enormous gravity*; at least not for ever. The earlier it is admitted and recognized, the less regretful it is going to be, intellectually. We are going to see, with proofs, that the 'greatest blunder' is in fact the failure to realize the role and the importance of the matter in the Universe and its gravity in determining dynamics and the physics of relativity.

4.4 Gravity of Our Universe

As I already emphasized, the theories of relativity and dynamics were considered complete before we acquired any significant knowledge about the real Universe we live in, where we construct and test our theories and where all the physical phenomena happen. Therefore, these theories naturally assume, as their basis, the *Minkowski space-time, which remains isotropic and homogeneous in every inertial frame* (Einstein, 1956). Then, there is no privileged reference frame to define the true motion and there is no physical sense in the notion of absolute rest. But the space of the factual Universe is not empty; there is no place in the Universe we can find as empty space, devoid of matter and gravity. We know that gravity affects and modifies spatial and temporal intervals, and affects the general features of dynamics of particles. Yet, we explicitly assume a Universe with empty space, devoid of gravity, in our fundamental theories. There is an enormous conceptual and quantitative difference between the physical phenomena in empty space and those in the gravitational potentials of vast amounts of matter. Hence, it is obvious that we deemed our theories complete too early to incorporate in them the one decisive factor that controls and determines relativistic physics and dynamics – the gravity of the matter in the Universe, or 'cosmic gravity'. Worse still, we neglected to reconsider the basis of these theories for a century, even after an accumulating body of empirical evidence with warning clues became available.

How much is the gravitational potential (g-potential) generated by all the matter in the Universe? In my experience, most people answer (without a calculation) that the g-potential is negligible because most of the matter is very far. The g-potential of the Earth in my office is $\phi_e = -GM_e/R_e \simeq 10^8\ m^2/s^2$. Now we go and pick a source far away, whose pull of gravity we do not feel, like the Sun, and calculate the potential, $\phi_s = -GM_s/R_s \simeq 10^9\ m^2/s^2$. That is 10 times larger! If one calculates the potential of the entire mass in the Milky Way, we get a value that is a thousand times larger. The trend is clear – distant masses dominate. This should not be surprising because the g-potential of a mass varies as $1/r$, whereas the amount of matter and its mass increase as the volume, r^3. Then, the total g-potential increases with the extent of the matter distribution, as r^2; that is how the distant matter dominates in its contribution to the g-potential.

We can venture to calculate the g-potential of all the matter in the Universe approximately, with an average density 10^{-26} kg/m^3 and a spatial extent that is approximately the Hubble radius $c/H_0 \approx 10^{26}$ m (the estimated 'size' of the Universe). Since the age of the Universe is finite, the potential is also finite. The size of the Universe might be much larger than what we can see today, but what is important for the calculation is only the matter up to a distance $d = cT_u$, where T_u is the age of the Universe. The matter outside this 'sphere of influence' is not yet visible, and its gravity is not yet felt here. Calculating an approximate value of the Newtonian gravitational potential due to the matter and energy in the Universe with the Hubble size is simple (Fig. 4.1).

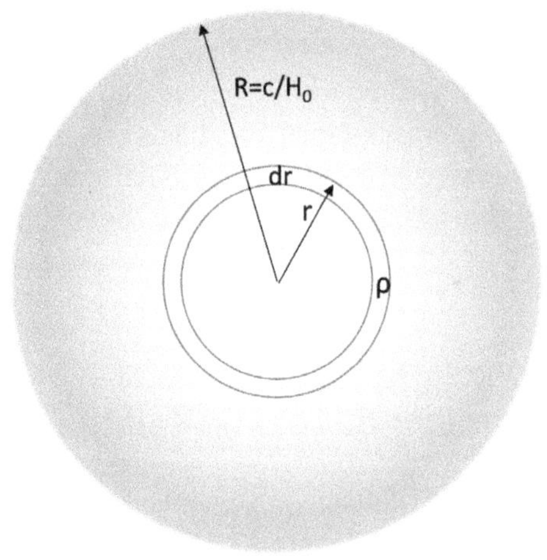

Figure 4.1 Approximate calculation of the Newtonian gravitational potential of the matter and energy in the Universe, by adding up the potentials due to the spherical shells with the uniform density ρ and thickness dr. The integration is from $r = 0$ to the distance that is causally connected to the observer, which is determined by the age T_u of the Universe, $r \simeq cT_u$.

The magnitude of the potential due to the annular volume is $d\phi = (4\pi r^2 dr \times G\rho)/r$. Integrating, we get

$$\Phi_u \simeq \int_0^{cT_u} 4\pi G\rho r dr = 2\pi G\rho c^2 T_u^2 \quad (4.10)$$

The answer is the interesting number $\Phi_u \simeq 10^{17}$ m^2/s^2, which is about a billion times larger than the potential due to the Earth! Not only that it is very large and dominating, it is also numerically very close to the square of the speed of light! Yes, *the cosmic g-potential is everywhere equal to the square of the velocity of light,* $\Phi_u \simeq c^2$, for an unknown reason. We will discuss the implications later. With this result, we are justified in the assertion $\Phi_u/c^2 = 1$. This is remarkable. The thrill of new physics starts right here.

A reconsideration of the entire fundamental physics is obviously essential, demanded by both logical factors and empirical evidence. The gravitational background of a matter-filled Universe that can affect relativistic physical phenomena and dynamics is an undeniable reality. No physicist who adheres to logical and empirical rigour would continue on the naive belief that our existing theories based on the explicit assumption of negligible gravitational potentials in the physical environment could be correct and complete. A paradigm based on empty space could not be maintained or defended logically. The only possibility was obvious as soon as the gravitational nature of the Universe was realized. The gravitational Universe is the paradigm and the basis for the correct theory of relativity and dynamics. That was the genesis of my theory of Cosmic Relativity.

4.5 The Paradigm of Cosmic Relativity

The paradigm of Cosmic Relativity rightfully replaces Einstein's Special Theory of Relativity (STR) because of the simple reason that the Universe is not empty, contrary to the fundamental and essential assumption of the STR. Cosmic Relativity modifies the basis of both the fundamental theory of relativity and the theory of dynamics in many significant ways. We have a totally gravitational paradigm as the basis of relativity and dynamics. Since the physical reason is the gravity of the cosmic matter, Einstein's General Theory of Relativity (GTR) needs a major modification, being a theory of gravity. The unavoidable and perennial background gravitational fields of the cosmic matter should be the irreducible part of Einstein's equations of the theory of gravity. The energy-momentum of the matter in the Universe is included in the new 'Centenary Einstein's Equations' of the modified GTR (Unnikrishnan, 2018a). This modification takes away the feature of the 'general coordinate invariance' from the GTR, due to the physical reality of the unique master frame of the Universe. The theory becomes fully 'Machian'; the well-known enigma of the inertial forces – pseudo-forces without sources – is solved and these forces naturally emerge as true gravitational forces, with the cosmic matter as their source (Unnikrishnan, 2014).

After Hubble's discovery of the recessing galaxies (Hubble, 1929), the Universe that emerged was really vast, with enough matter and its gravity to define the space and time for physics. Though gravity was a well-understood interaction, neither Einstein nor anybody else reconsidered, in the light of this crucial knowledge of cosmology, the course already taken by the theories of relativity. The current theories of relativity continue to assume that our physical space is the non-existent empty space, and our time has no relation to the gravity of the cosmic matter. It is as if the standard physics of today considered the Universe we see as a huge make-believe film set with virtual reality galaxies and matter devoid of gravity; these theories continue in their course set definitely in the 1920s. In reality, there is one master frame with its material reference markers as the galaxies, and one universal time, given by the expanding Universe and its evolutionary clock. This is also termed 'absolute time', but only in the sense of being defined by the single master reference frame of the material Universe. 'Universal time' is the appropriate term. This is distinctly different from Newton's ideal and metaphysical absolute time in the empty space.

The hypothetical ether was stationary and without evolution, whereas the real Universe is evolving, as evident in its expansion. Thus, *the evolving Universe defines a universal time*, which the ether was devoid of. The empty space is homogeneous and isotropic for all inertial observers, and the measures of spatial and temporal intervals or the 'metric' remain homogeneous and isotropic in

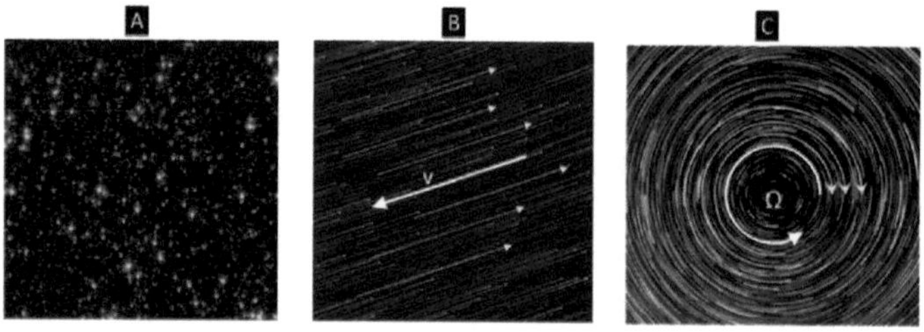

Figure 4.2 Motional anisotropy in the Universe. (A) In a frame at rest relative to the matter distribution, the Universe is homogeneous and isotropic on large scales. (B) In an inertial frame moving at the velocity v, there is an enormous matter current in the opposite direction (yellow arrows). (C) In a non-inertial frame with rotation at the rate Ω, the matter current is 'rotational', resulting in a large gravitomagnetic field.

every moving frame. The invariant isotropy is a property specific to the Lorentz Transformations (LT) wherein the metric $\eta_{ij} = diag\{-1, 1, 1, 1\}$ remains invariant under motion and the LT. (We use the symbol η_{ij} for the Minkowski metric and g_{ij} for the general metric, following the standard convention.) This is the very basis of the Special Theory of Relativity. *The STR needs both the empty space and the LT in its structure, to be consistent with its postulates.*

The matter distribution in the Universe is more or less homogeneous and isotropic on very large scales, but only for an observer who is 'at rest' relative to the average matter distribution. It is clear that moving in the matter-filled Universe generates a relative flow of matter opposite to the motion of the observer, and this *current of the gravitational charge* should generate a large gravitomagnetic field and consequent physical effects akin to the magnetic effects in electromagnetism (Fig. 4.2). Space becomes anisotropic with this flow and the metric changes, reflecting this anisotropy. The same anisotropy is accurately reflected in the Doppler anisotropy of the temperature of the Cosmic Microwave Background Radiation (CMBR).

The Doppler dipole anisotropy of the temperature of the CMBR enables a fairly precise determination of one's motion relative to the cosmic frame (Section 3.4). Conceptually very significant is the fact that the monotonically decreasing temperature of the CMBR provides us with a universal absolute clock and time, synchronized everywhere in the Universe to a precision better than a part in 10^6. This demolishes the arguments for the lack of universal simultaneity of spatially separated events. The CMBR provides a universal (absolute) reference for real motion as well as a universal time. However, it is the physical fact of the *matter-energy in the Universe being the source of dominant gravity* that determines the physics of relativity and dynamics. This fact defined and guided the theory of Cosmic Relativity, which theoretically complemented several decisive experimental results, and completed the gravitational paradigm of relativity and dynamics. Some details are discussed in subsequent sections.

As soon as we acknowledge that our Universe is filled with matter and energy, the spatial anisotropy due to the current of the cosmic matter in any moving frame and its gravitational effects cannot be denied. The Universe is then the master frame, relative to which every motion is well defined. This anisotropy will then reflect in the metric description of frames that are moving relative to the distributed matter in the Universe as well. However, we saw that an observer at rest has an isotropic metric. The transformation to a moving frame *generates* an anisotropic metric. The Lorentz Transformations cannot do this! The LT preserve the isotropy of the Minkowski metric intact, because the Minkowski metric remains invariant under the LT. It is a straightforward logical and mathematical deduction, given the factual Universe with matter at the observed critical density, that the Lorentz Transformations and the invariant Minkowski space-time structure should be abandoned. The real Universe is maximally 'Lorentz violating', contrary to the general belief. Then, Einstein's STR also is to be abandoned. It cannot be the correct theory operating in a Universe with vast amounts of matter and its gravity. Not only the total loss of its mathematical basis, but also the factual existence of an absolute reference frame for motion, with a well-defined absolute time, dethrone the STR for good. In fact, the unjustified and widely held belief of 'faith' is the only factor that delayed the past opportunities to correct the misguided course of the physics of relativity and dynamics.

The anchoring realization of the theory of Cosmic Relativity is that all physical phenomena, their experience and tests, as well as their understanding through the theoretical frameworks that we construct, happen in this single factual Universe, in the perennial and irreducible presence of all the matter and its gravity. We do not have the freedom to assume any postulate that is in conflict with this factual situation.

Let us review the main features and physical consequences of the Cosmic Relativity paradigm. The details of Cosmic Relativity, as the complete theory of dynamics and relativity, is the topic of another monograph in preparation (Unnikrishnan, 2022). Here, we will briefly explain the salient features and then plunge into the relation between cosmic gravity and time, which is the topic of our current discussion.

Since the matter in the Universe is distributed approximately uniformly on a large scale, without appreciable coherent motions on such scales, the cosmic gravitational potential has nearly the same value everywhere in the Universe. The small variations due to an isolated galaxy or star, in their vicinity, are less than a part in a thousand to a million. *As well known in fundamental physics, a potential without spatial or temporal variations cannot be detected by any physical means.* Not that there are no physical changes due to such an interaction potential, but those changes are identical in every physical system. Therefore, there is no way to compare with a reference system and detect the physical effects. If we move at a constant velocity through the homogeneous and isotropic matter distribution, the potentials are different. In particular, there is now a matter current relative to our moving frame and the associated 'gravitomagnetic vector potential' that we discussed earlier. However, these are also without variations in space and time, and therefore, undetectable by any physical means. The physical consequence of nearly uniformly distributed matter is the highly homogeneous gravitational potentials, which remain so for every observer who is in uniform motion. *The fact that there are no locally measurable physical effects of a constant potential field is indeed the Principle of Relativity (PoR)*! We have discovered the deepest principle underlying dynamics, as a consequence of the gravitational Universe. We do not need to postulate the PoR. It emerged naturally from the observed and verified properties of the single Universe we inhabit. The PoR is strongly tied to the observed homogeneity of the matter-energy distribution and its gravitational potentials.

We found that the Lorentz Transformations and the invariant Minkowski space-time structure of the empty space are inconsistent in the space of the matter-filled Universe. If the LT are inconsistent, how do we describe the motion relative to the master frame of the Universe mathematically? Surprisingly, the Galilean Transformations (GT) correctly give us, along with the observed anisotropy of space with matter, the most important relativistic feature of motion – time dilation! I will show this using the actual FLRW metric, with the very slow time evolution of expansion. Then $a(t) \simeq 1 + Ht = 1 + t/\tau$, where $\tau = 1/H$. τ is of the order of the age of the Universe. The line element for an observer at 'absolute' rest is

$$ds^2 = -c^2dt^2 + (1 + Hdt)^2(dx^2 + dy^2 + dz^2) \tag{4.11}$$

The GT are $x' = x - vt$ and $t' = t$. Choosing the direction of motion along the x-axis, at the velocity $v(t)$, we have the transformed line element (the y and z coordinates are suppressed for brevity),

$$\begin{aligned} ds'^2 &= -c^2dt'^2 + (1 + Hdt')^2dx'^2 & (4.12)\\ &= -c^2dt^2 + (1 + Hdt)^2(dx^2 + v^2dt^2 - 2vdxdt) \\ &\simeq -c^2(1 - v^2/c^2)dt^2 - 2vdxdt + dx^2 + 2Hdtdx^2 & (4.13) \end{aligned}$$

I have neglected the smaller terms with both H and v, since we consider $v \ll c$. Denoting cdt as $d\tau$, the first term is to be equated to $g_{00}d\tau^2$, and the second term is $(g_{01} + g_{10})dxd\tau$. We see the miraculous appearance of the (squared) time dilation factor $g_{00} = (1 - v^2/c^2)$, in the Galilean transformed line element! Also, the anisotropy terms are present as $g_{01} = g_{10} = -v/c$. The coefficient of the last term, Hdt, is irrelevant for laboratory physics, because it is essentially dt/T_u, where T_u is the age of the Universe.

Thus, the GT to a frame with the velocity v relative to the master cosmic frame transform the metric of (t, x) is

$$\begin{bmatrix} g_{00} = -c^2 & g_{01} = 0 \\ g_{10} = 0 & g_{11} = 1 \end{bmatrix} \quad \xrightarrow{GT(v)} \quad \begin{bmatrix} -c^2(1 - v^2/c^2) & -v/c \\ -v/c & 1 \end{bmatrix} \tag{4.14}$$

Non-zero g_{i0}, the gravitomagnetic potential, gives the observed dipole anisotropy. The physical temporal duration, as measured by a clock in the moving frame, is now $\sqrt{-g'_{00}}dt' = (1 - v^2/c^2)^{1/2}dt$. The GT indeed contain the physics of time dilation with v as the *absolute velocity*, or the velocity relative to the cosmic frame. It is very important to note that there is no restriction on the motion to be inertial in Cosmic Relativity. Thus, Cosmic Relativity releases the theory of relativity from the artificial straight jacket of inertial motion right at the outset.

In Cosmic Relativity, the motion relative to the Universe is 'absolute', with its measure given by the dipole anisotropy. Therefore, it is obvious that there is no ambiguity as to which observers are really moving and which are at rest. All the paradoxes of relativity, debated and discussed by generations, disappear without a whimper from the theory, for ever, when Cosmic Relativity takes the place of the STR! Only those clocks that move relative to the matter distribution have a time dilation, given accurately by the factor $\int dt\sqrt{1 - v^2(t)/c^2}$. We will soon delve into the real physical reason for the motional time dilation.

Since the cosmic frame is the master reference frame for all motion, with the Galilean features for the transformations to moving frames, Cosmic Relativity has a remarkable and daring prediction for the nature of the propagation of light. *Cosmic Relativity predicts that the relative velocity of light is Galilean*! Relative to an observer moving at the velocity v, light propagating along the same direction will have the relative speed $c - v$. Similarly, light propagating in the opposite direction will travel at the relative speed $c + v$. Note that this is the velocity relative to an observer. The speed of light relative to the cosmic frame is always c. The situation is kinematically similar to the one with the hypothetical ether. If Poincaré went a step further and speculated that the ether is in fact the Universe and its gravity, the course of physics would have been very different. The definite and spectacular experimental evidence for the Galilean nature of the propagation of light is discussed in Chapter 6.

The gravitomagnetic potential represented by the metric term g_{0i} is very important in many ways, apart from quantifying the anisotropy of space in a moving frame. It is very similar to the vector potential $\vec{A}$ in electromagnetism. Thus, its 'curl' is the gravitomagnetic field, $\nabla \times \vec{A}_g = \vec{B}_g$. Since $\vec{A}_g = -\vec{v}/c$ for the cosmic gravitomagnetic potential, we get

$$\vec{B}_{gU} = -\nabla \times \vec{v} = -2\vec{\Omega} \tag{4.15}$$

Ω is the angular velocity of a reference frame, rotating in the matter-filled Universe (Fig. 4.2C).

We should naturally wonder why there is no sign of the gravitational constant G and the matter density of the Universe ρ_u in any of our expressions representing the gravitational potentials. The answer reveals why we have not realized the role of the gravity of the Universe for so long, despite many obvious clues. To see the answer clearly, we go back to the earlier calculation of the Newtonian gravitational potential of the matter and energy in the Universe. For the spatially flat Universe ($k = 0$) with the critical density $\rho_c = 3H_0^2/8\pi G$, we have

$$\Phi_u = 2\pi G\rho c^2 T_u^2 = 2\pi G\rho c^2/H_0^2 \simeq c^2 \tag{4.16}$$

(The age of the Universe is approximately the inverse of the Hubble parameter $T_u \simeq 1/H_0$.)

Not only that the cosmic gravitational potential is equal to c^2, it remains so always, when the average density of matter and energy in the Universe has the standard evolution. Though the density decreases due to the steady expansion, a larger volume of the Universe and its gravity come into the picture progressively (Ryden, 2002). The result is an invariant Newtonian potential. This is a well-known fact and a compelling clue, but ignored hitherto as irrelevant for physics, unfortunately. The equality $\Phi_u = c^2$ requires that the Universe is spatially flat ($k = 0$) and the factual average density be equal to the critical density. In other words, the property of the Universe being curvature-less in space is time independent – once flat, always flat. Then, the ratio $\Phi_u/c^2 = 1$ remains invariant (time-independent) at the value 'unity'. The presence of the constant '1' in any expression is invisible! That is the reason why this prefactor that contains the physical parameters of G and ρ is not readily visible in any expression of dynamics and relativity. The factor $g_{00} = 1$ in a metric of relativity is just this factor. The gravitomagnetic potential $-v/c$ is really $-v\Phi_u/c^2$. *Suppose I make the phenomenological claim that the speed of light has nothing to do with relativity, and the*

Lorentz factor in relativity really is $\gamma = (1 - v^2/\Phi_u)^{-1/2}$ *and not* $(1 - v^2/c^2)^{-1/2}$. *A theory of relativity constructed on such a claim would agree with every empirical fact and experimental result known*! Therefore, it is very clear that we took the wrong turn in 1905. For the time being, we will keep the factor as c^2, with the understanding that it does not imply any link with the propagation of light. That is, while the propagation of light is controlled by the cosmic gravitational potential, the physics of relativity has no direct relation to the propagation of light. Light and electrodynamics were essential tools for studying relativity, but in reality, the relativistic effects were gravitational. However, in an era when only rudimentary knowledge existed about the factual Universe, nobody could connect the physical link.

The study of the cosmic gravitational potentials in moving frames will lead us to several remarkable results. The potential Φ_u changes to $\Phi_u'(v) = \gamma\Phi_u = \Phi_u/\sqrt{1 - v^2/c^2}$ in a moving frame. Concurrently, the gravitomagnetic potential $-v/c$ appears, because of the matter current in a moving frame (Fig. 4.2). Note that

$$\Phi_u'^2 - \vec{A}_g^2 = \gamma^2\Phi_u^2 - v^2\gamma^2\Phi_u^2/c^2$$
$$= \frac{\Phi_u^2}{1 - v^2/c^2} - \frac{v^2}{c^2}\frac{\Phi_u^2}{1 - v^2/c^2} = \Phi_u^2 \qquad (4.17)$$

Several important results that provide the ultimate understanding of the meaning and the content of some key foundational principles of physics of the past 400 years follow naturally and effortlessly. *Cosmic Relativity eliminates the dominance of postulates and 'laws' in fundamental physics, and derives the core results of dynamics and relativity as the physical consequences of the gravity of the matter in the factual Universe* (Unnikrishnan, 2009, 2022).

4.6 Some Results from Cosmic Relativity

(1) **The Principle of Relativity**: We already saw that the Principle of Relativity (PoR), which is the central principle of modern relativistic physics, emerges as a physical consequence of the gravitational paradigm of Cosmic Relativity and dynamics. The PoR is the consequence of the impossibility to sense

uniform gravitational potentials of the nearly homogeneous and isotropic distribution of matter in the Universe, during inertial motion.

(2) **The Law of Motion (Newton's Law)**: Physics treats Newton's laws of motion as a set of postulates. However, in Cosmic Relativity, everything in relativity and dynamics is a consequence of the cosmic gravity. Consider a frame in arbitrary motion through the cosmos. I write the relevant component of the gravitomagnetic potential arising from motion along the 'i-th axis',

$$g_{i0} = A_i = -\frac{v_i \Phi_u}{c}\left(1 - v^2/c^2\right)^{-1/2} = -\gamma \frac{v_i \Phi_u}{c} \tag{4.18}$$

Though g_{i0} is homogeneous in a uniformly moving frame, which implies the principle of relativity, g_{i0} *becomes time dependent if there is an acceleration*, since v becomes time dependent. *The physical effect of a time dependent potential is a reactive force on the accelerated system*,

$$F_i = m_g \frac{dA_i}{cdt} = -\frac{m_g \Phi_u}{c^2}\left(\gamma a_i + \gamma^3 v_i (\vec{v} \cdot \vec{a})\right) \tag{4.19}$$

We see that accelerating a body requires overcoming this cosmic gravitomagnetic reaction and hence a force F_i proportional to the mass, which is *the full relativistic form of Newton's law of dynamics*. The factor $\Phi_u/c^2 = 1$. We have derived Newton's law of motion as a consequence of the cosmic gravity! All dynamics is entirely determined by cosmic gravity.

The conceptual implications are unprecedented. *I have unified dynamics and gravity*. Truly fitting to the fact that both the law of motion as well as the law of gravity were discovered by the same person, Newton's dynamics is contained in Newton's gravity. Who would have imagined that Newton's law of motion was a gravitational law, and had the same feature as the Lenz's law in electromagnetic reactive induction!

(3) **The Equivalence of the Inertia and the Gravitational Mass**: In Indian mythology there is the impressive story of the popular god Krishna offering his mother the grand view of all the worlds, by opening his little mouth, while an infant. That is the moment when she realizes the supreme dimension in him,

representing all the worlds. Cosmic Relativity offers the grand view of the deep unification, right at the foundations of physics, of principles and physical effects by the single fact of cosmic gravity.

Look carefully at the expression for Newton's law of motion.

$$\vec{F} \simeq \left(\frac{m_g \Phi_u}{c^2}\right) \vec{a} = m_i \vec{a} \tag{4.20}$$

What we call the inertia or the inertial mass in dynamics is the product of the gravitational mass and the gravitational potential of the Universe. $m_i = m_g \Phi_u / c^2$. It is a hybrid quantity, arising solely from the interaction of a body with all the matter in the Universe. Hence the ratio m_i / m_g is universal. We have derived the Equivalence Principle of Galileo and Newton, as a simple consequence of the gravity of the Universe. (Unnikrishnan, 2009, 2014). What Einstein had to postulate in the beginning of his road to the GTR, I have derived from the first principles and the factual properties of the Universe, in which the principle is operative. There are very important implications to our finding.

(4) **The Tests of the Equivalence Principle**: Newton's law of dynamics and the equivalence principle have the same physical content and one implies the other, through cosmic gravity. This means that a test of Newton's law of motion is the same as a test of the Weak Equivalence Principle (WEP). The implication to modern experiments to test the WEP are discussed in the reference (Unnikrishnan, 2014).

(5) **The Motional Time Dilation**: Cosmic Relativity establishes beyond doubt that the time dilation of a moving clock is factually the gravitational time dilation due to the cosmic matter. All physical processes happen in the perennial presence of the cosmic matter and its gravity, and the rates of these processes have their 'normal' value (say, f) at absolute rest, in the gravitational potential $\Phi_u = c^2$. Any motion relative to the cosmic frame alters the potential to a higher magnitude that changes the rates of the time evolution to a lower value f'. The reason is simple: a body moving along an arbitrary trajectory at the speed $v(t)$ experiences the deeper potential $\Phi' = \gamma \Phi_u$ (see eq. 4.17). Thus, the gravitational potential energy of a moving body increases in magnitude by the factor $m(\Phi'_u / \Phi_u) =$

$m/\sqrt{1-v^2/c^2}$. *The resulting increase in the inertia is the reason for the slowing down of every physical process.* The increased inertia, relative to the inertia at rest, is $m_i'/m_i = \Phi'/\Phi = 1/\sqrt{1-v^2/c^2}$. Then, the modified rate of the process, relative to the rate at rest, is in the inverse ratio, $f'/f = \sqrt{1-v^2/c^2}$ (see the next chapter, Sections 5.2–5.6 for many details.).

Clocks themselves are physical processes, with an intrinsic 'rate'. Therefore, the moving clock will have a gravitational time dilation, due to its decreased rate of the underlying physical process. The time T' marked by a moving clock (relative to any stationary clock) is

$$T'/T = f'/f = \sqrt{1-v^2/c^2} \tag{4.21}$$

and the time dilation of a moving clocks is

$$\frac{\delta T}{T} = \frac{T'-T}{T} = \sqrt{1-v^2/c^2} - 1 \simeq -\frac{v^2}{2c^2} \tag{4.22}$$

What we thought as the time dilation 'due to the relative motion in empty space' is factually the gravitational time dilation due to the gravity of the matter in the Universe! The time dilation of a clock depends on its 'absolute' velocity (speed) of motion – the velocity of true motion, relative to the master frame of the Universe. Note that this is consistent with the term g_{00} in the Galilean metric in a moving frame, which determines the time dilation of the clocks (eq. 4.14). This is a revealing discovery that provides a complete understanding of the time dilation of moving clocks. *There is only one kind of time dilation, and that is the gravitational time dilation.* Hitherto, physics was familiar only with the gravitational time dilation due to local sources of gravity, like the Earth. Now, Cosmic Relativity makes it clear that what was misunderstood as the time dilation 'due to relative motion', is really the gravitational time dilation due to the gravity of the entire matter in the Universe, induced by real motion relative to the cosmic frame. Quantitatively, this is the most accurate description of time dilation, applicable to any situation involving arbitrary motion. Also, the scope for any kind of paradox about time dilation has disappeared completely.

(6) **The Origin of the Inertial Forces – The Coriolis Force**: A frame like our Earth is rotating in the presence of the matter

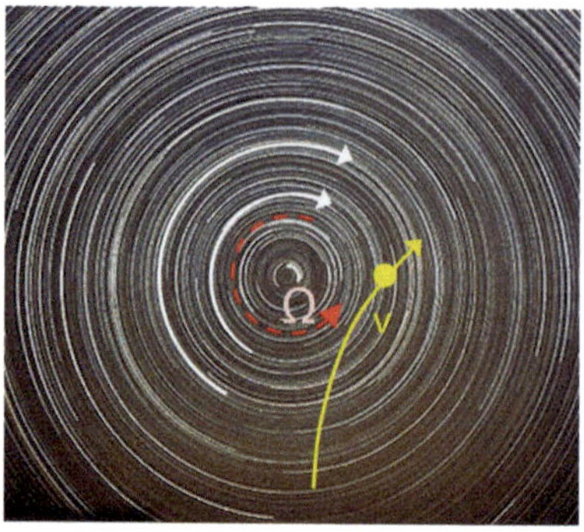

Figure 4.3 The Coriolis deflection of a body moving at the velocity v in a rotating frame in the matter-filled Universe. The rotation of the frame at the rate Ω is counter-clockwise here (red dashed arrow). The cosmic matter current is clockwise (white arrows). The gravitational force between this current of matter and another moving body (yellow) deflects it.

of the Universe. This means that relative to us, all the matter is rotating as an enormous whirlpool of the current of the charge of gravity (Fig. 4.3). There are physical consequences, as one can imagine.

In a rotating reference frame, the gravitomagnetic vector potential has the non-zero curl,

$$-\nabla \times \vec{v} = -2\vec{\Omega} = \vec{B}_g \tag{4.23}$$

This is the cosmic gravitomagnetic field in a rotating frame. This affects the motion of objects in a rotating laboratory. An object moving at velocity u will feel a gravitomagnetic force $F_{gm}(u) = m_g\vec{u} \times \vec{B}_g = -m_g u \times 2\Omega$. This is indeed the Coriolis force.

We have found that one of the familiar inertial forces, remained elusive as a pseudo-force for three centuries, is the gravitomagnetic force arising because of the rotation relative to the matter distribution in the Universe. (The phenomenon of the Coriolis force was known for a long time, though not to Newton. The description and the analysis of the force in rotating frames were done by Gustave de Coriolis in 1835.) We see that the Coriolis force $m_g\vec{u} \times \vec{B}_g$ is just the gravitational Lorentz force, similar to the magnetic force $e\vec{v} \times \vec{B}$ that deflects a moving charged particle. In spite of the uncanny similarity between the dynamical and electromagnetic phenomena, and in their mathematical descriptions, one did not make the obvious connection earlier.

The true cause of such forces can be traced deeper, as the force between two currents – between the current defined by the motion of the body and the current of the external matter. Hence, it is exactly similar to the forces between electric currents, studied by A. Ampère. In electromagnetism, parallel currents attract and opposite currents repel. In gravity, the converse happens, consistent with the contrasting behaviour of the charges in the two interactions. This has important applications, for example in the analysis of the physical factors in tropical cyclonic storms.

Can we also derive and explain the more common centrifugal force from cosmic gravity?

(7) **The Origin of the Inertial Forces – The Centrifugal Force**: The centrifugal force was Newton's argument for the reality of the absolute space. Ernst Mach demolished Newton's reasoning, pointing out that the rotation was not relative to the empty space, but relative to other masses and the distant stars. Therefore, the centrifugal force might be originating in some interaction with the rest of the matter in the Universe. The Machian speculation was profound, but never taken to its logical end on the basis of modern observational evidence. Besides, Einstein's rejection of an absolute frame and his insistence on the general coordinate invariance, buried the revered Mach's principle.

For motion in a curved trajectory at a constant speed, the direction of the gravitomagnetic potential $\hat{A}$ is changing at the rate Ω, even though its magnitude A is a constant. Any change in the potential is a force acting on the body. The change in the direction of the cosmic gravitomagnetic potential is opposite to the change in the direction of the movement of the frame. Hence, *for circular motion, the force is outwards, irrespective of the sense of motion*. From the time dependence of the direction of A;

$$F_c = m_g \frac{d\vec{A}}{dt} = m_g A \frac{d\hat{A}}{dt} = m_g v\Omega = mv^2/r. \qquad (4.24)$$

Figure 4.4 The gravity of the Universe helps Einstein to balance the turning bicycle against the Earth's gravity, even with a slant that would make him fall otherwise. The balancing force comes due to the gravitational interaction with the matter in the Universe.

I have solved the long-standing and perplexing problem of the inertial pseudo-forces (or fictitious forces) that troubled Newton, Mach and Einstein (Fig. 4.4). They are genuine gravitational forces with the cosmic matter as the source. They are not orphan 'fictitious forces'; they arise because the inertia of a body is factually the product of the mass of the body and the gravitational potential of all the matter in the Universe. It is particularly satisfying that this gravitational solution is final, requiring no revision, because it agrees completely with what is observed in the laboratory and what is observed as matter and energy in the Universe. It is truly amazing that the distant part of the Universe is more important in determining the inertia, than the nearby Universe, because of its larger volume and matter content; the farthest 1/3 of the Universe has 4 times the amount of matter than the nearer 2/3, contributing more than half the total gravitational potential. This deep link with the entire Universe is what we experience as the real forces in our journeys in the winding paths of life.

But, I must stress *the profound difference between the gravitational force that we feel from the Earth and the inertial gravitational force we feel in an accelerating or rotating cabin.* The pull of gravity is an 'electric' type force, directly from the charge of gravity – the mass. The latter is inductive, from the time varying gravitomagnetic potential, analogous to the 'induced electric force' in electromagnetism, $F_i = -\partial A_i/\partial t$. It exists only when one accelerates relative to the cosmic frame. It is not the same as the gravitational force of attraction $F_g = -\nabla\phi_g$ between two masses that makes objects fall and keeps the Moon in its orbit. Put compactly, 'inertia' is the gravitational *response* to the accelerated motion of the 'mass'; they are distinct in character. This difference between gravity and the forces felt in acceleration suggests that Einstein's Equivalence Principle is violated in the case of the gravitational time dilation of clocks: *there will not be any relative time dilation of clocks in an accelerating frame*, but such a test is yet to be done. We discussed the problem of the time dilation of accelerated clocks in some detail in the previous chapter. Now we understand completely why there cannot be any *relative time dilation* for such clocks, apart from the differential motional time dilation that happens in the process of bringing them together for comparison. However, there is the gravitational shift of the frequency of light in an accelerated frame, because of the Doppler shift and the Galilean nature of the propagation of light.

We do not discuss here the numerous other new results obtained in Cosmic Relativity, especially those pertaining to electrodynamics and quantum mechanics. Fascinating and important they are for sure, but not directly relevant for our discussion of gravity and time. The details of the theory of Cosmic Relativity, its experimental basis and tests, as well as its many results and applications are discussed in the references cited (Unnikrishnan, 2009, 2022).

We conclude this section with a thought experiment that provides a significant insight. A hypothetical situation in familiar electrodynamics can help to dissolve away any residual doubt as to the verity of the gravitational paradigm, and convince oneself of the results of Cosmic Relativity that we just discussed.

Imagine a situation where there is a slight charge asymmetry of less than 1 part in 10^{35} between the proton and electron, and that the atoms are slightly charged. Such a minuscule asymmetry cannot be detected in laboratory experiments. In the pre-cosmology era, while Maxwell was completing his theory, no knowledge about the matter content of the Universe was known and such a tiny charge asymmetry in matter known then (solar system, Milky Way, and a few nebulae around us) would not have made any difference to the observed features of electrodynamics. However, the true reality would have been the enormously charged Universe with a huge electric potential generated by all the unseen charged matter, but *undetectable in all inertial experiments* because constant potentials cannot be locally detected. Therefore, the PoR would have been valid. However, in situations of accelerated motion of a charge, strange extra forces would appear without any identifiable source because the potentials become time dependent, due to the time dependent velocity. Charges in rotation would experience pseudo-magnetic forces; they would deflect in the rotating laboratory without any source of a magnetic field in the lab. A magnetic moment in a rotating frame would feel an anomalous torque, without any identifiable source. *The situation would have required postulating pseudo-forces in electrodynamics, proportional to the electric charge.* One would be faced with a new electromagnetic inertia, in addition to the usual inertia proportional to mass, forcing one to ascribe it to the special motion relative to the 'absolute' space, as Newton did to discuss his 'water in a rotating pail' experiment. The vexing puzzle would not be solved until somebody realized, from alert wisdom or actual observations, that there indeed was an enormous amount of matter in the Universe with a slight charge asymmetry. We do not see such anomalous electromagnetic forces; the entire electrodynamics is accounted for with local sources only because the matter in the Universe is essentially neutral. *In sharp contrast, the Universe is gravitationally charged, with the mass and energy of the cosmic matter as the enormous charge of gravity.*

This reiterates that we are grossly wrong in ignoring the gravity of the matter around us in our fundamental theories of dynamics. That the conventional pseudo-forces could indeed be from cosmic gravity had been argued by D. Sciama (Sciama, 1953), and occasionally by some others. However, stopping there gives an incomplete and inconsistent story, rejected several times in the past by the physics community. One has to realize that such a framework also demands a preferred frame and the notion of absolute rest, and hence, the compatible Galilean notions of space, time, and the propagation of light. Even a partial implementation of the Machian paradigm necessarily rejects Einstein's Special Relativity. *One cannot have the Machian paradigm without a privileged reference frame for motion and the Galilean propagation of light.* We are at once led to a natural paradigm where the Universe is the master frame (instead of the hypothetical ether) and its universal gravitation is the determining interaction (instead of electromagnetism with its limited scope). Cosmic Relativity demands and predicts that the relative velocity of light and gravitational waves are Galilean, and all relativistic effects are factually gravitational effects. Dynamics and its laws are then dictated by cosmic gravity.

I have promised to discuss the experimental results that prove the Galilean nature of the propagation of light, with the speed $c \pm v$, relative to an observer moving at the velocity $\pm v$. This is described in some detail in Chapter 6. Thus, the cardinal prediction of Cosmic Relativity is verified in a reliable experiment. The same experiment falsifies the fundamental postulate of Einstein's STR, of the invariance of the relative velocity of light.

The General Theory of Relativity itself has to be reconstructed in a way that retains the agreement with all its impressive precision tests done during a century and yet compatible and *consistent with the prior and unique background Universe* with so much matter and its gravity. A major prediction of the GTR was gravitational waves. Because of its anchor on the STR, gravitational waves are supposed to propagate like light in the GTR. However, since light is already shown to be Galilean, from direct experiments, gravitational waves also are Galilean, as evident from the recent simultaneous detection of both gamma rays and gravitational waves from a distant astrophysical event (Abbott, 2017).

Einstein's equation for gravity are modified by including the cosmic gravity and the master frame as part of the equation itself. The resulting 'Centenary Einstein's equation' is

$$R_{ik} - \frac{1}{2}g_{ik}R - \frac{8\pi G}{c^4}T_{ik(U)} = \frac{8\pi G}{c^4}T_{ik} \tag{4.25}$$

The extra piece $T_{ik(U)}$ on the left of the equation is the energy momentum tensor of the Universe, included as the non-removable integral part of the equation itself (Unnikrishnan, 2018a). Adding this term incorporates the Universe as a preferred frame for Einstein's equations, making the theory maximally 'Machian'. This breaks the characteristic feature of the general coordinate invariance assumed in the GTR, consistent with the lack of such invariance in the factual Universe. Then, Mach is naturally integrated with Galileo, Newton, Lorentz, Poincaré, and Einstein, retaining only the consistent elements that constitute a satisfactory completion.

4.7 The Cosmic Clock

I mentioned the fact that the temperature of the CMBR in the Universe is the same everywhere to a great accuracy (better than a part in a million), and that it is slowly decreasing, with its present value at 2.72548 ± 0.00057 (above absolute zero, about $-270.425°$C). The second feature of the monotonic decrease makes it equivalent to a clock that keeps time everywhere in the Universe. The first feature shows that it is a clock that is synchronized everywhere in the Universe - an absolute time and a reference for absolute simultaneity. In fact, the FLRW line element that we discussed as the basis for theoretical relativistic cosmology,

$$ds^2 = -c^2dt^2 + a^2(t)\left(\frac{1}{1-kr^2}dr^2 + r^2d\theta^2 + r^2sin^2\theta d\phi^2\right) \quad (4.26)$$

shows this universal absolute time explicitly in the first term, as *the time coordinate t and its metric that is independent of the location or the epoch*. This represents our Universe, with $k \simeq 0$, and yet, we deny such a universal time in our physics! Since we live in this Universe, the only one accessible to us, it is plainly clear that *our physics is not logically consistent with our cosmology*. This is a very troublesome situation. It is as if we are somewhere outside this Universe, in empty space, studying various physical systems, and the Universe is just one of the physical systems for our study, with equal status as a stone or an electron that we study. But, our theories as well as all our experiments and observations are within this Universe, in its mighty and real gravitational presence. Let us not ever forget that.

We cannot use the CMBR for our terrestrial timing applications, because the change in the temperature is extremely slow, a millionth of a degree in 3000 years; nevertheless, it is a genuine clock that marks events on the scale of cosmic epochs. What is more important for physics is to realize that this cosmic clock holds the same synchronized time across galaxies separated by millions of light years. Therefore, prescriptions like the one we saw in the Lorentz Transformation $t' = \gamma(t - vx/c^2)$, by which two clocks that are separated by 10 million light years ($x \simeq 10^{22}$ m) and moving together at 100 km/s, would have a difference of 10000 years, is not consistent with cosmology. *The LT and the theory of relativity built on it, which is the STR, are not consistent in our matter-filled Universe.*

Coming back to terrestrial matter and our familiar clocks, we are now fully prepared to discuss and understand the modification of time due to motion. The discussion of cosmic gravity and its physical effects has made it very clear that mere relative motion does not cause any time dilation. The time dilation of moving clocks is the gravitational time dilation of clocks that move relative to the matter in the Universe. We will soon be able to answer every question about motion and time dilation that remained puzzling and without any convincing answers, in the prevailing theories. We are finally on the clear ground of the physics of time dilation, free of paradoxes, and firm on logic and reason.

4.8 Summary

In this chapter, we examined the features of our 'once-given' dynamical Universe, its contents and its gravity, as observed and understood. The realization that the Universe contains vast amounts of matter, more or less uniformly distributed, implied the obvious fact that its enormous gravity can affect dynamics and the progress of time. We learned how the matter and radiation in the Universe serve as the reliable master reference frame for gauging motion, and how the slowly changing temperature of the radiation serves as an absolute universal clock that is synchronized everywhere in the Universe. A simple calculation sensitized us to the amazing and important fact that the gravitational potential of the matter and energy in the Universe is equal to the square of the velocity of light. Then we explored some relevant results of the theory of Cosmic Relativity, the cosmic gravitational paradigm for relativity and dynamics. The consideration of the gravitational potentials in frames that are moving relative to the matter in the Universe yielded many fundamental results: the principle of relativity, motional time dilation, Newton's law of motion and dynamics, the origin of inertia, the equivalence of mass and inertia, and the recognition and derivation of the centrifugal and Coriolis 'fictitious forces' as genuine gravitational forces with the cosmic matter as their source. With the insights thus gained about relativity and dynamics, and in particular about motional time dilation, we can now examine the modifications of time in motion with unprecedented clarity and precision, without ever stumbling on the slightest of ambiguities.

Chapter 5

Gravity's Time

We are starting this chapter as people enriched and strengthened with reason, nurtured by logic and empirical evidence, and with a deep insight. The physical world is comprehensible when we pay attention to its integrated existence and interconnections. With our understanding of the gravitational Universe, even in the simplest terms, we can go a long way, almost the complete way, in understanding the physical time and its modification. We will not dwell on the past history, when for decades many compelling clues for the cosmic gravitational effects were ignored by a conservative physics community. We will focus now on our goal of understanding gravity's time.

5.1 Gravitational Time Dilation

The Special Theory of Relativity dictates that two clocks in relative motion have a relative time dilation. It does not identify unambiguously the clock that runs slower than the other; instead it connects the time dilation to the *apparent kinematic motion*, because relative inertial motion cannot be unambiguously ascribed to one or the other frame. Clearly, two clocks at relative rest do not have a

Gravity's Time
C. S. Unnikrishnan

ISBN 978-981-4877-72-5 (Hardcover), 978-1-003-25656-4 (eBook)
www.jennystanford.com

relative time dilation, as long as they are in an inertial frame. What happens when the frame is non-inertial?

What does it mean to be non-inertial? Well, since current physics does not know the origin and reason for inertia, it cannot answer this question satisfactorily. But, non-inertial motion and acceleration are closely related, almost synonymous. The only conceptual advance made in this context was Einstein's use of the equivalence of mass and inertia to postulate the Principle of Equivalence between gravity and acceleration. This enabled him to predict the phenomenon of gravitational time dilation. We have already seen the relevant derivation (Chapter 2).

Two clocks A and B that are separated in height in a gravitational field would have a relative time dilation given by

$$\Delta T/T = (T_B - T_A)/T \simeq gd/c^2 \tag{5.1}$$

with the upper clock at B running faster (recording more elapsed time). More precisely, the rates of the two clocks are proportional to the local gravitational potential $\phi(r)$. This follows from the GTR, where the physical time T is related to the coordinate time t by the relation $dT = \sqrt{-g_{00}}dt$. Near the Earth or similar spherical massive bodies, $g_{00} = -(1 - 2GM/c^2r)$. Since the quantity $2GM/r \ll c^2$, we have the relation

$$\sqrt{-g_{00}} = \sqrt{(1 - 2GM/c^2r)} \simeq (1 - GM/rc^2) = (1 + \phi(r)/c^2) \tag{5.2}$$

Therefore, the gravitational modification of time is given by $\delta T/T = -GM/rc^2 = \phi/c^2$. A clock runs slower, indicating less time, in the gravitational potential. When two clocks in different gravitational potentials are compared after a duration T in the gravitational field, they would have accumulated a *relative difference* in their recorded durations,

$$(T_B - T_A)/T = (\phi_B - \phi_A)/c^2 = \delta\phi/c^2 \tag{5.3}$$

This can be written approximately in terms of the gravitational field (equal to local gravitational acceleration) $g = -d\phi/dr$, when the distance between the clocks $\delta r = d$ is small enough to treat the field as uniform,

$$(T_B - T_A)/T \simeq \frac{d\phi}{dr}\frac{\delta r}{c^2} = gd/c^2 \tag{5.4}$$

When the separation between the clocks is comparable or larger than r, the more accurate expression with the values for the local gravitational potentials needs to be used. We will see an example when we discuss the clocks in the satellites of the GPS.

There is a subtle issue in the comparison of separated clocks; very subtle indeed. We are talking about two clocks that are spatially separated, one at r and another at $r + dr$. What do we really mean by the relative time dilation of spatially separated clocks? We cannot compare two clocks that are at two different places. They need to be brought to the same location for comparison, and that involves movement. No analysis ignoring this factor of the moving of the clocks before their comparison is rigorous and complete. However, the motion of a clock is always in the cosmic frame, and not in the non-existent empty space of the STR. This makes a huge difference for the comparison of clocks in accelerated frames, as we already saw in Section 3.13.2.

5.2 Why Is There a Motional Time Dilation?

We briefly discussed in the previous chapter how a moving clock in this Universe experiences a different gravitational potential that slows down its rate. We will discuss the motional time dilation again, now in the context of the real experiments and their interpretations that happened in the past one century. Then, we will see the true vulnerability of the grand edifice of physics we simply believe in.

The relevant questions have been asked by generations of people who came across the time dilation of moving clocks. The questions remained without consistent answers, to be repeated again and again, not because people could not understand the simple formulae or the mathematical structure of the theory, but because the answers were never logically satisfactory. The very fact that the answers in the standard (advanced or otherwise) textbooks disagree with each other, and even with Einstein's own answer on the twin-paradox, is ample indication of the general confusion. Now we are more than equipped to decisively answer the question why and how exactly the motional time dilation happens. If the time dilation is a real physical effect, it must have a physical reason that is equally real. To start with, the motion that is linked to the time dilation should be real. A clock that suffers a change in its rate should be really moving. But, what is 'real movement'?

The motion of a clock can be sensed only relative to some reference markers. But, there is no way to distinguish between the inertial motion of the clock and the motion of the markers. *How can there be a difference in the rate of a clock just because there is a relative motion between the clock and some reference markers?*

A bit of alert contemplation will tell us that the physical reason behind the time dilation must be some difference in the physical environment of the clock. The relative motion between the environment and the clock must have changed the physical interaction between the environment and the clock. *Then the true reason is the difference in the physical interaction, and the motion is only an agent that effects this difference.* At once, *we see that the relation between the references of motion and the clock can only be gravitational*; gravitation is the only universal interaction that can affect any kind of clock, and it acts over a long range, effective as far as the knowable Universe. Therefore, *what is important is not the motion relative to some reference markers, but the motion relative to some massive markers that have gravity*! Then the motion makes a difference to the strength of the interaction, because the physical potentials of interactions, both of gravitation and electromagnetism, depend on the relative velocity.

Let me say this in a different way. Irrespective of the theory of time, and its nature, the factual observations of the modifications of time during motion (or in a gravitational field) are done with material clocks that work by counting repetitive physical processes. Hence, a physical process has to be affected, for the rate of the clock to be affected. This is possible only if there is a physical interaction that can affect the particular physical process. While the GTR insists that matter can affect space and time, no theory would say that matter-less space and time could affect any physical process. Therefore, the modifications of the rate of physical processes during motion must be linked to a physical interaction, and the only interaction that universally affects physical processes and all kinds of clocks, without exception, is gravitation.

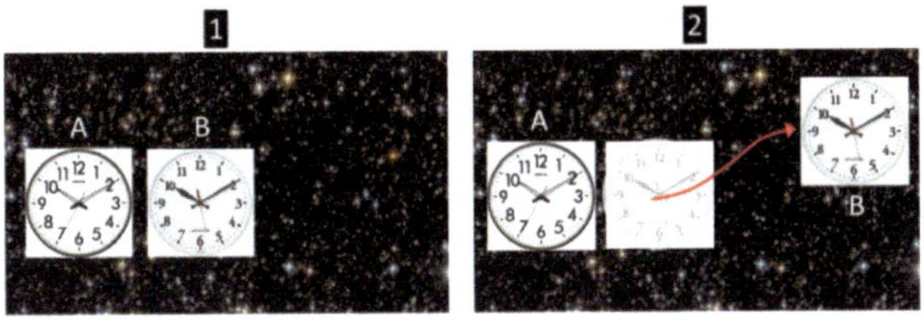

Figure 5.1 The gravitational potential experienced by a clock at rest in the Universe is different from the one experienced by a clock in motion. (1) Clocks A and B at rest relative to the cosmic frame experience the same potential Φ_u and have the same temporal history. (2) The clock B is in motion at the speed $v(t)$. The potential experienced *during the motion* is larger, $\Phi_u' = \gamma\Phi_u$ and B has the gravitational time dilation relative to A, $(T_B - T_A)/T = \Phi_u(\gamma - 1) \simeq -v^2/2c^2$.

5.3 Cosmic Gravity, Clocks, and Time

I will briefly repeat and then expand on the relevant material from the previous chapter (Section 4.5) that explained how exactly the gravity of the matter in the Universe affects the rate of moving clocks. The gravitational potential of the entire matter in the Universe is denoted Φ_u. We calculated it approximately and found that its magnitude is $\Phi_u \simeq c^2$ (the sign of the gravitational potential is negative, relative to the situation of no gravity). This is consistent with the metric component $g_{00} = -1$, with the factor c^2 absorbed in the definition of the time coordinate. Henceforth, we set $\Phi_u = -c^2$.

Two clocks A and B at rest in this cosmic potential will have the same gravitational effect on their dynamics (Fig. 5.1). The time they indicate is already dressed by cosmic gravity, but it is undetectable because the gravitational effect on both the clocks is identical! As E. Mach famously wrote, the Universe is not 'twice given', one with the factual matter distribution and another with only 'space' devoid of any matter. We cannot do any experiment to compare with the hypothetical situation in which the matter in the Universe is removed. All what one can do is to see the relative time dilation when one clock is compared with another in a *different* gravitational potential. A difference in the gravitational potential can arise either due to the presence of a local mass nearer to one of the clocks, or due to the difference in the motion of the clocks relative to the matter distribution in the cosmic frame.

The clock A that is at rest in the Universe experiences the gravitational potential Φ_u and the clock B that is moving in an arbitrary trajectory at the speed $v(t)$ experiences the deeper potential $\Phi'_u = \gamma\Phi_u$. Therefore, the clocks have different rates, determined by the gravitational potentials they experience. I showed in Section 4.6 that the cosmic gravitational time dilation of the moving clock B, relative to the stationary clock A, is

$$\frac{\delta T}{T} = \frac{T' - T}{T} = \sqrt{1 - v^2/c^2} - 1 \simeq -\frac{v^2}{2c^2} \qquad (5.5)$$

The last step assumes that $v \ll c$. The velocity that appears in the factor γ is the 'absolute' velocity – the velocity of true motion relative to the cosmic frame (CF). It has a unique value at each moment, unlike the relative velocity of the STR, which changes with the choice of the reference frame. Note that this formula for the time dilation looks superficially similar to Einstein's formula for the time dilation in the STR. However, its content is very different – it is the cosmic gravitational time dilation during the arbitrary (inertial or otherwise) motion of the clock, at the *absolute* velocity $v(t)$. The velocity $v(t)$ does not depend on the choice of any inertial reference frame; it is always relative to the unique cosmic frame. (In the STR, v is the inertial relative velocity, which depends on the choice of the reference frame.) The motional time dilation is a direct and inevitable consequence of the fact that the Universe contains matter with gravity. This gravitational time dilation would be absent in empty space without matter and gravity. Therefore, the time dilation of a moving clock is indeed because of the physical interaction of gravity. But we did not realize this fact, because the source of gravity in this case is the large amount of matter that is distributed thinly in our vast Universe. *There is only one kind of time dilation, and that is the gravitational time dilation.* Quantitatively, this is the most accurate description of the time dilation, applicable to any situation involving any arbitrary motion.

Clearly, the time dilation of the STR, supposedly due to the relative motion in empty space, has no place in physics, or in this Universe, because the vast amount of distributed matter in the Universe is an observed fact. Then, we have to accept its gravity and its physical effects. *The cosmic gravitational time dilation accounts for the whole observed time dilation of elementary particles, etc.* This discovery is enough reason for the shift to the cosmic gravitational paradigm of Cosmic Relativity.

5.4 Finally, an Obituary for the Twin-Paradox

Having seen that the motional time dilation depends on the velocity relative to the unique and 'absolute' master frame of the Universe, the STR and its physics based on mere relative velocities in the gravity-less and non-existent empty space are to be abandoned for ever. With the theory, its paradoxes go as well. The vast literature on the twin-clocks problem in relativity is littered with many illogical arguments that cite entirely different physical reasons to explain the single fact of the observed asymmetrical time dilation between the two clocks. If one examines a dozen standard textbooks on Special Relativity, one finds at least three or four different physical reasons for the single result of the asymmetrical relative time dilation of two clocks that were separated at some relative velocity, and then were brought together for a direct comparison; some talk about a shift in the 'line of simultaneity' when one of the observers reverses the path, others rely on the Doppler effect of the transmitted signals between the clocks (though real experiments may not have any such signalling), and some others cite the 'acceleration' as the cause, without mentioning how exactly it affects the clocks. In fact, many are unaware that if acceleration is invoked as a factor in calculating the time dilation, the problem shifts in scope from the STR to the GTR, because the underlying physical fact is the postulated equivalence of acceleration and gravity; an acceleration by itself cannot affect clocks.

Many texts say that the asymmetric answer to the twin-clocks problem is 'obvious', because the time measured by a clock depends on its path in space-time. But, one forgets that the path in the symmetric theory of the STR depends on the choice of the rest frame. If A is chosen as the rest frame, the space-time path of B is shorter. But, if B is chosen as the rest frame, the space-time path of A is shorter! Almost all the books claim that the problem is solved entirely within the STR. In particular, these authors would vehemently oppose any suggestion for invoking the GTR or gravity to resolve the paradox. Obviously, they do not realize that they are dismissing Einstein's own answer to the twin-paradox, without justification. (When I pointed out Einstein's gravitational resolution to an author, who discussed the twin-paradox in a popular book on time and insisted on a resolution within the STR, the response was 'Einstein is not God'. On the other hand, my Cosmic Relativity paradigm for relativity and time dilation was not acceptable to the same person, because 'Einstein couldn't be wrong'!) Some authors go as much as mentioning gravity, as equivalent to the acceleration, but skip the simple calculation as 'beyond the scope of the book'. Many authors are not aware that the problem can be posed with three observers (clocks), without any acceleration (see Fig. 3.16). Thus, entirely different physical reasons are advanced as the correct solution for the twin-clocks paradox in the STR, because all of them result in the identical mathematical expression for the relative time dilation. But, as we saw in detail (Section 3.12), none of them offers the correct solution.

If both observers are accelerated, then people are at a loss in dealing with the problem because there is no easy way of picking one as the privileged inertial observer. Worse, all real experiments have non-inertial observers and accelerated clocks! The situation is surprisingly confused and messy.

The thousands of pages of incorrect reasoning that are written and published on the twin-clocks paradox can be, and should be, replaced by the correct statement, expressible in one sentence: *the reason for the asymmetric time dilation is the difference in the gravitational potential experienced by the clocks moving at different velocities relative to the matter in the Universe.* Since the matter in the Universe is visible for everybody, this reason cannot be denied and a simple calculation will show the perfect match between the reason and the observations. The gravitational reason is applicable for general motion, without any distinction between inertial and non-inertial motion.

Once this is verified, the STR stands falsified because one cannot have both the gravity of the cosmic matter and the STR posed as the reason for the time dilation; that would be double counting. But, the matter in the Universe is already observed and cannot be denied, and therefore its gravity is real. We who know, on every occasion, which clock will have the larger time dilation, and its real reason, should be at equanimity.

5.5 Gravity and Clocks

After fixing our ideas and conclusions about the intimate connection between gravity and time, it is insightful to examine how exactly gravity could affect specific kinds of clocks. This is important because if gravity affected (or not) each type of clock differently, depending on the mechanism of the clock, the connection between time and gravity would not be universal. We need to concretely see the universal influence of gravity on every kind of clock, from the mechanical ones with pendulum, spring or vibrating quartz, to the atomic clocks, and then even the 'living clocks' with their organic 'life-time'. If we can be convinced about the universality of the gravitational time dilation, we will truly understand the gravitational nature of time and its modification in a gravitational field.

The General Theory of Relativity does not ever address the physical reason why clocks suffer a modification of their rate in a gravitational field. In Einstein's relativity theories, the physical effects are addressed by examining the quantity representing the four-dimensional line element,

$$ds^2 = g_{ij}dx^i dx^j \tag{5.6}$$

The quantities g_{ij}, which depend on the coordinates, are the components of the metric that determine the modification of spatial and temporal intervals. In the simpler case of independent (orthogonal) Cartesian coordinates of space and time this is

$$ds^2 = g_{00}dt^2 + g_{11}dx^2 + g_{22}dy^2 + g_{33}dz^2 \tag{5.7}$$

The component g_{00} determines the modification of time. Since the presence of a gravitational field as well as changes in the coordinate systems are represented by the details of the metric, no further analysis in terms of the physical interaction is attempted in the GTR.

We can and should do better, because a better physical understanding is possible. To this end, let us see how a physical interaction like electromagnetism can affect a clock. An electric field or a magnetic field can affect a clock mechanism only if the mechanism involves the motion of charged particles. We call it an EM-clock. We will take an electron in a magnetic field, executing the circular cyclotron motion, as our oscillator for the EM-clock. With our expertise on clock making by now, we can arrange the rest easily, like a sensor to count the number of revolutions, the sensors and electronic feedback to maintain a constant magnetic field, etc. The cyclotron frequency for a particle of mass m in a magnetic field B is

$$\omega_c = \frac{eB}{m} \tag{5.8}$$

This is convenient for a clock, since the frequency does not depend on the velocity of the charged particle. We can get about 1 GHz as the oscillator frequency for an electron with 100 Gauss (0.01 Tesla) field. A skilled experimental physicist can design such a clock and operate it with a stability of 10^5, at least for short durations; not a great clock, but good enough for our discussion. The EM-clock is susceptible to both electric and magnetic fields, and needs to be shielded well. A small change in the magnetic field of 0.01 Gauss will amount to a change in the oscillator frequency by a part in 10^{-4}, or 0.1 MHz. Then it will drift away from a good quartz clock by a millisecond in 10 s. So, we see clearly how an interaction can affect the rate of a clock. In this example, we had a special situation of the mechanism of the clock being derived from the magnetic interaction and the perturbation also was chosen as magnetic. However, electric fields also will shift the frequency of the charged oscillator, by perturbing the dynamics and distorting the trajectory.

Even the rate of a pendulum clock can be affected by a magnetic field, because most materials have a response to the magnetic field and there could be also magnetic impurities in the non-magnetic material of the pendulum.

The rate of a quartz clock depends on the size of the crystal. Quartz has a thermal expansion coefficient of 10^{-5}/K. So, a degree change will alter its oscillation frequency by a similar fraction. The thermal effect can be radiative heating or even internal dissipation.

This is a common example of how interactions can change the rate of a clock.

However, none of these is fundamental and universal. Such perturbing effects on clocks can be mitigated either by better isolation or by changing the mechanism. But, these examples clearly show that the rates of clocks are subject to long range interactions. *If the effect of the interaction is universal, affecting identically every clock and physical process conceivable, then we say that the time itself is affected by the interaction.* What is universal in different kinds of clocks is what is called 'time'.

There is only one interaction that couples to everything in the Universe without exception and that is gravitation. Besides, gravity cannot be shielded (Unnikrishnan and Gillies, 2002). Gravity couples to mass and to the mass-equivalent of energy ($m_E = E/c^2$). Therefore, it affects every physical process, altering its energy by $\delta E = m\phi_g$ where ϕ is the gravitational potential. What are the consequences for a clock in which the mechanism uses some physical process like the oscillations of a pendulum, or the vibrations of a quartz crystal, or the transitions in an atom?

5.5.1 Time Dilation of Atomic Clocks

Let us take the atomic clock first. The clock is based on some transition between two energy levels with the difference $E_t = E_2 - E_1 = hf_c$, where h is the Planck's constant and f_c is the frequency of radiation that can cause the transition 'at resonance'. The frequency of the resonant transition in the atom is used as a stable reference for the much smaller frequency of the oscillations of a quartz crystal, in the 'atomic clock' (see Section 1.9 for some details). This frequency is geared down to the 'one pulse per second' standard time. Since the energy in the atom also is subject to gravity, just as the mass of the atom, putting the atom in a gravitational field implies a modification of its energy. The transition energy E_t is modified due to gravity by $E_t' = E_t - (E_t/c^2)\phi_1 = hf_c(1 - |\phi_1|/c^2)$, where ϕ_1 is the gravitational potential at the location (say '1') of the atom. We see that the clock frequency in the gravitational field is lower than the pristine frequency of the clock in the absence of the field. A slower oscillator translates to slower seconds and slower time (recall that the unit 'second' in a clock is reached by counting a fixed number of the cycles of an oscillator). Note that the modification of the oscillator frequency is proportional to the gravitational potential ϕ, and not to the strength of the gravitational field (pull), $g = -d\phi/dr$. To be clear, near a source like the Earth, the potential is $\phi(r) = -GM/r$ and the field is $g(r) = -GM/r^2$. Even in an idealized 'constant' gravitational field g, the values of the potential at two different points along the field direction are different, because $g = -d\phi/dr$. The frequency of another clock at the location '2', at a different potential ϕ_2, will be modified by $-hf_c\phi_2/c^2$. Then two clocks at the positions '1' and '2', with separation d in a constant gravitational field g, will have a gravitational potential difference

$$\phi_2 - \phi_1 = -d\frac{d\phi}{dr} = gd \tag{5.9}$$

Therefore, the difference in the rate of the clocks is $\Delta f_c = f_c(\phi_2 - \phi_1)/c^2 = f_c gd/c^2$. The clock in the deeper potential (closer to the source) ticks slower, its 'seconds' being longer than those of the clock in the weaker potential. Thus, we get the *relative gravitational time dilation between two clocks* in a constant gravitational field,

$$\frac{\Delta T}{T} = gd/c^2 \tag{5.10}$$

The fractional time dilation does not depend on the properties of the atomic clock. It is universal and applies equally to all atomic clocks, made with different atoms and different transitions.

5.5.2 Time Dilation of Mechanical Clocks

Next we consider 'mechanical' clocks, which rely on the actual motion of a mass, made periodic by using a spring or by the pulling action of gravity, as in a simple pendulum clock. How does the gravitational interaction modify the time marked by such clocks, by exactly the same amount as in atomic clocks? There are two contributions to the modification of the period of an oscillator in such clocks. One is of course the gravitational modification of the energy of the oscillations. The other is subtle, linked to a deep fact about mass and inertia.

First, we examine a mechanical oscillator constructed from a body of mass m attached to a spring with the 'spring constant'

k. Unlike a simple pendulum, such an oscillator does not need external forces like gravity for its functioning. If put in a gravitational field, the mass m, which is the charge of gravity, interacts with the gravitational field. The spring gets pulled a bit, but the mass continues to oscillate in its new position. The formula for the period of a spring-mass oscillator is $T = 2\pi\sqrt{m/k}$. *The subtle fact is that the m that appears in this formula is the 'inertia'*, also called just 'mass' conventionally. We have discussed the difference between the two earlier: m_g or the 'mass' is the charge of gravity, and the inertia is the resistance to motion, to be called the 'inertial mass' m_i. In the last chapter, we saw that the inertial mass is in fact the mass dressed by the cosmic gravitational potential, $m_i = m_g(\Phi_u/c^2)$. Inertia changes in the presence of other masses, due to the gravitational interaction. In fact, this was predicted by Einstein right after he formulated the GTR. After deriving the approximate equations of motion in his theory, Einstein wrote in the monograph *The Meaning of Relativity*,

> The inert mass is proportional to $1 + |\phi|/c^2$, and therefore increases when ponderable masses approach the test body.

(It is striking, and continues to surprise me, that he did not ever realize that the factor '1' in the expression for the 'inert mass' comes from the gravitational interaction with the matter in the Universe.) The mass m_g in the gravitational potential has the interaction energy $E_g = m_g\phi$. This energy is equivalent to a small increase in the inertia, $\delta m_i = |E_g|/c^2 = m_g\phi$. With the increased inertia, the period of the oscillator increases: $\delta T/T = (1/2)\delta m_i/m_i = m_g\phi/2m_ic^2$. Since $m_g/m_i = 1$, in the absence of the local extra potential, $\delta T/T = \phi/2c^2$. This is (only) half the gravitational time dilation derived for the atomic clocks.

There another equal contribution, from the change in the average energy of the oscillations, is similar to the change in the energy of the atomic levels. The energy in the mechanical oscillations is $E = mA^2\omega^2/2$. Note that the energy is proportional to the square of the frequency, unlike the linear relation $E = h\nu$ for the atomic systems. This is equivalent to a mass E/c^2, which has the gravitational interaction energy $\delta E = -E\phi/c^2$. Since $E \propto \omega^2$, we have $\delta E \propto 2\delta\omega$. That means $\delta\omega/\omega = \delta E/2E = -\phi/2c^2$. So, the decrease in the frequency due to the diminished energy of oscillations also contributes to the time dilation. The total gravitational time dilation is $\delta T/T = 2 \times \phi/2c^2 = \phi/c^2$. This applies for all clocks that rely on mechanical oscillators.

For a pendulum clock, the period depends on the length of the pendulum and the local gravitational field, $T = 2\pi\sqrt{(g/l)}$. However, the exact expression that distinguishes the gravitational mass m_g and the inertial mass m_i is

$$T = 2\pi\sqrt{m_i l/m_g g} \tag{5.11}$$

So, *'mass' appears in both its roles (as the gravitational charge and as the inertia) in a simple pendulum clock*. Though the ratio m_i/m_g is universal, the same for all bodies, the two quantities are not equal, $m_i \neq m_g$, because the inertia results from the gravitational interaction. In a local gravitational potential, the inertia increases as before, and results in half the time dilation. The other half comes from the diminished energy of the oscillations due to the gravitational potential, similar to the case of the spring oscillator. The energy in the oscillations is $E = mA^2\omega^2/2$, and the decrease in the frequency is $\delta\omega/\omega = \delta E/2E = -\phi/2c^2$. Then the total gravitational time dilation of a pendulum clock is $\delta T/T = 2 \times \phi/2c^2 = \phi/c^2$, which is the universal expression.

Just to avoid any confusion, I state the obvious. The gravitational time dilation of pendulum clocks is not because of any variation in the pulling force g of the Earth. (If g changes, as it occurs from place to place, time shown by the pendulum clock changes drastically. The rate of the clock can increase or decrease. But, that is not gravitational time dilation.) As we already saw, what is relevant for the time dilation of clocks is the difference in the gravitational potentials. We are considering two pendulum clocks at different potentials in a nearly 'uniform gravitational field'.

Similarly, in the clocks based on the electro-mechanical oscillations of the quartz crystals, there are two contributions to the gravitational time dilation, one from the change in inertia of the crystal body and the other from the gravitational interaction with the energy of oscillations.

What about the earth-clock that marks time by just rotating? Again, it is the modification of the inertia of the Earth due to the gravity of the Sun that is the reason for the gravitational time dilation of the earth-clock. The time marked by the rotating Earth slows down in the Sun's potential as $\delta T/T = \phi_s/c^2$. Since the potential ϕ_s depends on the distance from the Sun, the length of the day should decrease due to the gravitational time dilation by about 30 μs when the Earth is at the closest to the Sun in its orbit, compared to when it is at the farthest. But, all clocks on the Earth suffer the same time dilation and we do not (and cannot) notice or measure this.

We have explicitly demonstrated that the gravitational time dilation is universal, equally applicable to all kinds of clocks. My demonstration shows that time is just the physical processes of matter in evolution, and *the only reason why gravity can modify time is the gravitational interaction of matter with matter.* Only matter and its energy have gravitational interaction, and gravitational time dilation is possible only because time is identical to the count of the physical processes in material clocks. I went into the minute details of gravitational time dilation for the different kinds of clocks to convince everybody that there is no time other than the time marked by physical clocks, as evolving physical processes.

Now we can freely explore the motional time dilation, understood as the cosmic gravitational time dilation experienced in motion.

5.6 Time and the 'Absolute' Velocity

We are now in a new paradigm of motion where there is a well-defined physical meaning to 'absolute' motion and 'absolute' time, unlike in the current physics, which denies absolute motion and absolute time. The absolute or unique reference to motion is the Cosmic Frame (CF) defined by the average distribution of matter and energy in the Universe. The magnitude and direction of the motion of a body can be measured with the Universe as the master frame. The isotropic bath of the CMBR provides a convenient and precise way of measuring motion; this is identical to the motion relative to the distant stars, but the celestial sphere of stars is not a practical reference frame of precision. The slowly evolving and easily measurable temperature of the CMBR provides the measure of the universally synchronized absolute time. The main quantitative difference between the Cosmic Relativity paradigm and the STR paradigm is that in the Cosmic Relativity paradigm all relativistic effects depend on the absolute velocity, which is the velocity of motion relative to the cosmic frame. Then, only such real motion can be counted as physical. The relative motion due to the motion of a reference frame, which is a mere kinematical impression of motion, is not the true physical motion. Therefore, no physical effect can depend on such a relative motion of the STR.

The Lorentz–Poincaré theory of relativity also had the time dilation that depended on the 'absolute velocity', very much like the time dilation in Cosmic Relativity. However, the two are very different in concept and physical details. The modifications of duration in the Lorentz–Poincaré ether relativity was a function of the speed of the clock relative to the hypothetical stationary ether, very much like the motion in a fixed, but insensible, medium. With the ether as a privileged frame, the velocity is 'absolute'. In this sense there is a notional (theoretical) meaning to the real motion and its measure in the LPTR. However, *the fact that the hypothetical medium was unobservable by any means makes such an absolute speed unknowable*, which of course is the Principle of Relativity. Yet, it makes theoretical sense and the whole scheme connects real physical effects to the motion relative to the single privileged frame of the ether. In such a theory, only the clocks moving relative to the ether suffer the time dilation. With such an absolute criterion, there is no twin-clocks paradox and similar issues in the LPTR.

In contrast, the STR considers every inertial frame as equally privileged to be declared as the rest frame, or the stationary frame; the state of inertial motion is completely equivalent to the state of rest. Then there is no real motion. Motion in the STR is always relative to a frame arbitrarily chosen as stationary. The STR confers the privileged status to such 'rest frames'. *The clock in the frame chosen as stationary does not have any time dilation*, or any velocity-dependent relativistic effects, for that matter. Unlike the privileged clocks that are at rest in the ether, a stationary clock in the STR is the clock in any chosen rest frame, and then all other clocks that are relatively moving have the time dilation. With such a scheme, the time dilation in inertial motion cannot be a physical effect.

There are two reasons for this. One, with only relative velocity as physically relevant, the relative time dilation is a totally symmetric effect, as unequivocally stated by Einstein in his 1918 twin-paradox paper. Relative to a frame A that is chosen as stationary, a clock in a relatively moving frame B has a time dilation and relative to B, a clock in A has a time dilation. Then, the question 'which clock runs really slower?' is meaningless in the STR. Second, and perhaps more important, is that a relative velocity v can be partitioned in infinitely many ways as $v = v_A - v_B$! The rate of a clock moving at speed v_A (relative to some inertial frame F) is $dt'_A = dt(1 - v_A^2/c^2)^{1/2}$ and the rate of the clock moving at the speed v_B is $dt'_B = dt(1 - v_B^2/c^2)^{1/2}$. Assuming that the relative velocities are small compared to c, the difference in the time dilations of A and B (with $t_A \simeq t_B$) is

$$\frac{\Delta t'}{t'} = \frac{dt'_A}{t'_A} - \frac{dt'_B}{t'_B} \simeq \frac{-v_A^2}{2c^2} + \frac{v_B^2}{2c^2} \tag{5.12}$$

This is not equal to the relative time dilation when one of them (say, A) is arbitrarily chosen as the clock at rest:

$$\frac{\Delta t'}{t_A} \simeq \frac{-v^2}{2c^2} = \frac{(v_B - v_A)^2}{2c^2} \tag{5.13}$$

The theoretical structure of the STR, with the ambiguous relative velocity determining the real physical effects, is evidently fragile, if not inconsistent. In the cosmic gravitational paradigm, the relative time dilation is given unambiguously by the formula 5.12, where v_A and v_B are relative to the cosmic frame.

Though there were enough 'fixed' stars in the sky, Einstein understandably ignored them as a legitimate absolute reference frame for motion because they were mere markers without any physical significance, being sparse and very far away for any significant physical influence, as understood then. However, a reconsideration was definitely due after observational cosmology revealed the vast matter-filled Universe. But, this reconsideration did not happen till now, and that was indeed a tragedy.

We are now going to see the striking contrast between the new gravitational paradigm and the STR-based physics, in the face of the real experiments on the modification of time with precision clocks in motion. Having seen the full picture, as well as the logical and physical integrity of the gravitational paradigm, there is no prize for guessing which is the winning paradigm! After all, we are all equal partners of that paradigm, cohabiting in the same Universe and its gravity.

5.7 Time Asymmetry in Motion at Symmetric Relative Velocity

Suppose I have two synchronized atomic clocks in my laboratory, one of which is the reference clock A, which remains at rest in the lab. I move the other one B, through a short distance *d* at the speed *v*, relative to the lab. What can we say about the time in the moved clock? The unambiguous answer in the STR is that there is time dilation for B, and the time marked by B is given by

$$T_B = \sqrt{T_A^2 - d^2/c^2} = T_A\sqrt{1 - v^2/c^2} \tag{5.14}$$

An important characteristic of the time dilation in the STR based on relative velocities is that there is no dependence on the inertial speed of the reference frame. There is no need to ask the question whether the laboratory itself is moving. The clock in relative motion always runs slower, and 'ages' less. This means that two clocks synchronized with a third stationary reference clock in the laboratory (like an inertial space station), when transported in two different directions at identical relative speeds will have equal (isotropic) relative time dilations that are completely independent of the velocity of the laboratory. But, all experimental results known to date go against this characteristic trait of Einstein's theory! Unbelievable? We will scrutinize the results of the experiments and decide. Let us first examine the scenario and the relevant mathematical expressions, to prepare for the quantitative comparison.

In contrast to the STR, the time dilation in an absolute frame theory will depend not only on the velocity of the clocks relative to the reference clock in the laboratory, but also on the velocity of the laboratory relative to the cosmic frame (CF). Hence, the time dilation of the two clocks moving in opposite directions will be asymmetric, relative to the reference clock in the laboratory, in spite of their equal relative velocity. *What is most interesting, and damning for the STR, is that in relativity with an absolute frame, the clock that*

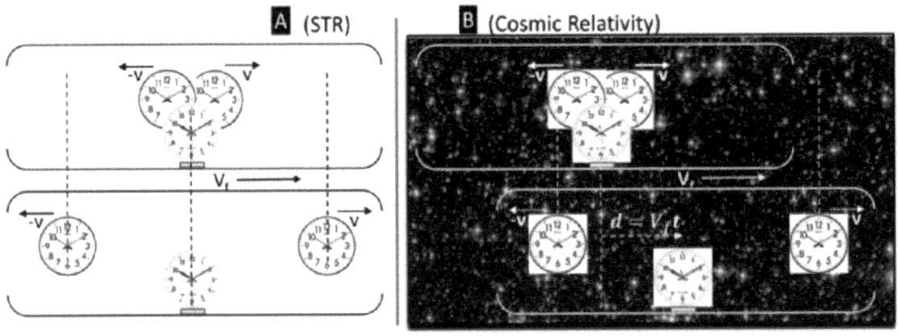

Figure 5.2 The crucial difference between the STR and Cosmic Relativity, for time dilation. The upper and lower panels indicate the situation at times 0 and t. (A) STR: the reference clock at rest in the lab has no time dilation. Two other clocks moving at the relative velocity $-v$ and v have equal time dilation. (B) Cosmic Relativity: all motion is relative to the cosmic frame (CF). The reference clock at rest in the lab is factually moving at the velocity V_f, and moves the distance $d = V_f t$ in duration t through the CF. Then the clock with relative velocity $-v$ has a factual velocity $V_f - v$, and *runs faster than the reference clock.*

moved relatively can run faster than a 'stationary' clock in the frame, and not slower as the STR demands. That should be shocking. If the space station is moving at velocity v_s in its orbit (relative to the CF) and the clock is moving out in the *opposite direction* at the velocity v_A relative to the space station, the clock is moving in the orbit at the real speed $v_A - v_s$ relative to the CF, which could be smaller in magnitude than $|v_s|$. Then the reference clock in the space station will have more time dilation than the 'moving' clock, as per Cosmic Relativity. *The Special Theory of Relativity never allows the possibility that a relatively moving clock runs faster.* This way, an experiment can easily decide and demarcate the correct theory.

This crucial difference is illustrated in Fig. 5.2. In the STR, the reference clock in the rest frame remains stationary, insensitive to any motion the frame may have relative to the stars. In contrast, in Cosmic Relativity, the reference clock itself is moving relative to the master frame of the Universe, at the velocity of the frame V_f. The clock moving at the *relative speed* v to the left is factually moving at the (absolute) speed $|v - V_f|$, which is smaller than V_f; therefore, that clock runs faster than the reference clock, and not slower. The clock moving at the relative speed v in the same direction as the motion of the reference frame, however, is factually moving at the speed $v + V_f$, and hence, will have a time dilation larger than what is predicted by the STR. The real shocker is when a clock in relative motion shows faster ageing. In the next section, we will see that direct experimental evidence for this existed for half a century, but the golden opportunity to correct the course of the physics of relativity was missed.

To stress again, in the STR, the time dilation is symmetric (isotropic) for clocks at equal relative speed v, $\delta T/T \simeq -v^2/2c^2$. The velocity of the rest-frame relative to any markers (like the distant stars) never appears in the expression for the relativistic effects in the STR. By contrast, in Cosmic Relativity, the time marked by the same clocks while the reference frame is moving at the velocity V_f (relative to the cosmic frame) is

$$\begin{aligned} T' &= T\left(1-(V_f \pm v)^2/c^2\right)^{1/2} \\ &= T\left(1-(V_f^2+v^2 \pm 2V_f v)/c^2\right)^{1/2} \end{aligned} \qquad (5.15)$$

Here, T is the notional cosmic clock that we discussed earlier. I have written the expression for the cases where V_f and v are collinear. Otherwise, the magnitude of $\vec{V}_f + \vec{v}$ should be used. The relative time dilation is

$$\begin{aligned} (T'-T)/T &= \left(1-(V_f^2+v^2 \pm 2V_f v)/c^2\right)^{1/2} - 1 \\ &\simeq -\frac{V_f^2}{2c^2} - \frac{v^2}{2c^2} \mp \frac{2V_f v}{2c^2} \end{aligned} \qquad (5.16)$$

The 'stationary' reference clock in the laboratory is factually moving relative to the cosmic frame at the velocity V_f. Hence it has the time dilation $\delta T/T \simeq -V_f^2/2c^2$. For two clocks A and B that share the velocity of the frame V_f (like a stationary reference clock A in a laboratory on the mobile Earth and another clock B in a train or flight), their relative time dilation is then

$$(T'_B - T'_A)/T' \simeq -\frac{v^2}{2c^2} \mp \frac{V_f v}{c^2} \qquad (5.17)$$

T' is the average duration of comparison. The second term renders the time dilation anisotropic or direction dependent! Note that it is irrelevant whether any of the motion – either of the clocks or of the reference frame – is inertial or not.

There is another important fact contained in the expression for the relative time dilation eq. 5.17. Since $vT' = L'$ is the distance travelled by the clock relative to the reference clock, the difference in the time recorded is $T'_B - T'_A = -T'v^2/2c^2 \mp V_f L'/c^2$. The last term is the same as the 'local time' in moving frames, introduced by H. A. Lorentz in his Lorentz Transformations. In fact, in the limit of very slow transport of the clocks, $v \to 0$, we have from the master equation for the time dilation eq. 5.16,

$$(T' - T) = dT' = T\frac{-V_f v/c^2}{(1 - V_f^2/c^2)^{1/2}} = -\gamma V_f L/c^2 \qquad (5.18)$$

which is the exact expression for the term in the LT. I have rederived the synchronization gap between two clocks, synchronized in proximity and then separated by slow transport to a distance $vT = L$. This is a reminder again to the fact that the local time vx/c^2 of the LT has no relation to the true relative velocity of light.

The difference between the time dilations in the STR and Cosmic Relativity is significant and easily measurable, but there is a subtle catch in trying to pick the correct theory using the prediction of clock rate asymmetry. Moving clocks break physical proximity, and they cannot any more be compared with the reference clock! To compare, either an electromagnetic signal needs to be sent back, or the clocks should be brought to the same location, reversing their velocity. Both methods cancel exactly whatever asymmetry accumulated in the outward trip, as we have already shown. So, the conventional two-way experiments will never reach the truth by distinguishing between the two theories.

There is one, and only one way out. *One needs a way of bringing back the clocks without reversing their motion and without changing their speed, ever.* This is possible by altering the direction of the motion incrementally or continuously, without changing the speed or the sense of motion (see the next section).

Now we examine a real experiment that was done half a century ago. Unfortunately for physics, the strong clues revealed then by the measured time dilation of the atomic clocks were missed or ignored.

5.8 The Hafele–Keating Experiment

In his 1905 paper on the STR, Einstein predicted that *a travelled clock transported in a round trip would gain time, or run slower relative to a clock that remained stationary in the same frame* (Einstein, 1905). In his words, 'if one of two synchronous clocks at A is moved in a closed curve with constant velocity until it returns to A, the journey lasting t seconds, then by the clock that has remained at rest the travelled clock on its arrival at A will be $\frac{1}{2}tv^2/c^2$ second slow'. The calculation of this effect in the STR is done using only the instantaneous *relative velocity* between the two clocks, and integrating (accumulating) the effect over the entire path. In any case, relatively moving clocks always have a time dilation in the STR. If the clock moved relatively, surely it has marked less duration.

We nearly learned a lesson on time dilation by discussing a flock of birds seen flying backwards, in a previous chapter. It is common to see an automobile or even an aircraft going backwards. At least in some situations, common sense alerts us to the fact that the impression is created because we are in a vehicle that goes faster than the one we are watching. However, the STR does not care for the physical reality that is evident to the common sense. The STR insists that the clock in our own frame, which by definition is the rest frame, is running at its proper rate and the clocks that are in relative motion always run slower. As mere physical objects in an 'objective' theory, the birds are in motion, while the observer is at *rest*, in the STR. To the question 'whose clock is going slower?', the theory gives the firm answer: the internal clocks of the birds are running slower, whether they are flying forward or backward. If they appear still, despite their flutter, they are also at rest and there is no time dilation.

Can I say that the birds that are 'really' moving forward slower than me, and hence appearing to be moving backwards, must also be ageing faster relative to me, guided by the common sense? Einstein's theory answers in the negative, because it is based on kinematics in the empty space. But, the *durations* cannot be compared unless I mark the time on the clocks at the same place twice, at the start and at the end; a duration is the difference between two simultaneities. The comparison of durations seems to require reversing the motion of the clocks. Therefore, there seems no way to verify my common sense guess for the birds that have flown past.

This by itself does not reveal any inconsistency in Einstein's theory in the most common situations, because an actual empirical comparison requires bringing the clocks together. Then the usual discussions reduce to a two-way transport of some clocks, where anisotropic (direction-dependent) effects cancel. However, the experiment can be done without reversing the motion of the clocks because the Earth is a globe and one can go around, as in a long flight, or (really) by slow transport in a ship. Then, the anisotropy of time dilation that is a telltale sign of the failure of the STR, as well as a strong support for relativity with a master frame, will be shown clearly.

It should have been indeed surprising, to say the least, if a clock comparison experiment ever revealed any evidence for a *transported clock running faster than a clock that remained stationary in the same frame*. If I synchronize two atomic clocks in the laboratory in Mumbai and then transport one to Paris by surface transport, what can be expected from the standard theory (STR) for the durations elapsed in the two clocks? A physical comparison is not possible, because the clocks are in two different locations. However, one can pose the question for a theoretical answer. Every physicist would agree that the transported clock records shorter duration because of the special relativistic time dilation while it was moved relatively to the stationary clock in the laboratory. We can continue with this experiment further, and move the clock to the east coast of USA and then to the west coast, and so on, in stages. In each leg of the trip, the clock is moving relative to the stationary laboratory clock. Eventually, the clock reaches back the laboratory in Mumbai. Now we have the opportunity for the direct physical comparison of the clocks (Fig. 5.3).

The result should have knocked every physicist out of one's chair and even out of one's office. *The transported clock runs faster and will be found to have aged more, instead of ageing less, contradicting the Special Theory of Relativity*! This could only happen if the clock 'really' moved slower than the laboratory clock. But there is no way for a transported clock to move slower than a stationary clock... or is there?!

An experiment was conducted in 1971 to measure the relative time dilation of a set of Cesium atomic clocks in motion, by carrying them in commercial flights around the globe, by J. C. Hafele and R. E. Keating of the US Naval Observatory (USNO) (Hafele and Keating, 1972). The 'Hafele–Keating experiment' compared two sets

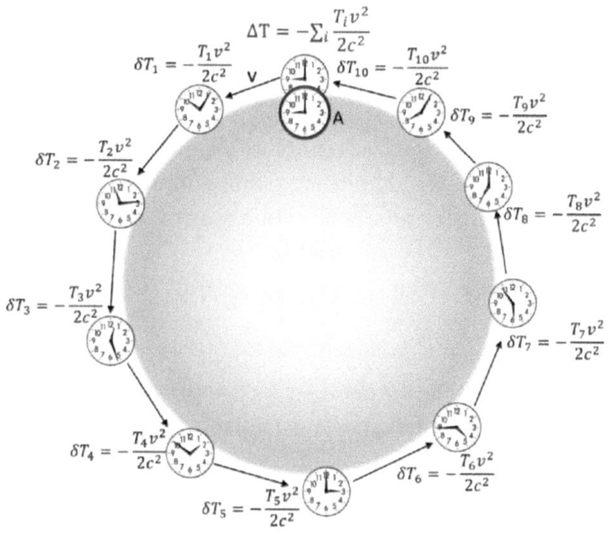

Figure 5.3 A clock in increments of inertial motion relative to a stationary clock (A) at the relative speed v should have the increments of time dilation $\delta T = -Tv^2/2c^2$ relative to A, if the STR is correct. When it reaches back after a round trip, the total time dilation should have been the sum of all incremental time dilations. But, experiments show a very different result. The observed time dilation relative to A depends on whether A has moved relative to a global frame, contradicting the STR.

of atomic clocks. One set of reference clocks was stationary in the Naval Observatory laboratory (equivalent to one of the twins who stayed back at home, in the 'twin-paradox'), and one set of four identically constructed Hewlett-Packard 5061A Cesium beam clocks was flown eastwards in commercial jet airplanes in a round trip (like the twin who set out on a round trip in a projectile). After completing the round trip and the comparison with the laboratory clocks, the same set was flown westwards (the adventurous twin taking his second trip), and then compared again after the round trip. The total flight times of these round trips were around 40–50 h. The average ground speed of the airplanes were around 220–240 m/s (about 830 km/h). The round trips (Washington→Washington) covered landing points in the northern hemisphere with latitudes ranging from 42°N (Rome) to 13°N (Guam), including Bombay (19°N and

Figure 5.4 The Hafele–Keating experiment on the motional and the gravitational modifications of time. The routes of the westward and eastward round trips are indicated. The ground speeds (speed relative to the reference clocks) and flight altitudes are approximately the same for the two trips.

Table 5.1 The flight parameters in the Hafele–Keating experiment (the trip duration includes the stoppage durations)

	Eastward flight	**Westward flight**
Trip duration	65.42 h	80.33 h
Duration in flight	$\simeq$ 40.2 h	$\simeq$ 48.6 h
Av. ground speed	243 m/s	218 m/s
Av. altitude	8.90 km	9.36 km
Av. latitude	34°N	31°N

Delhi (29°N), with an average latitude of about 32°N. The eastward and westward routes were slightly different; the eastward trip had fewer landing points and took less time (Fig. 5.4).

The flight parameters are summarized in Table 5.1. Since commercial flights were used, with many landing points, the calculation of the expected time dilation requires the actual flight data on the ground speed $v(t)$ and the height above the mean sea level $h(t)$ to be used. There are two contributions: the integrated gravitational time dilation with the gravitational potential estimated at the flight altitude, and the integrated motional time dilation with the speed of the airplanes relative to the laboratory clocks. In the STR, the laboratory clocks are at rest and the relative speed is just the ground speed of the aircraft.

The full data from the experiment, of the durations measured by the four clocks, had a large statistical spread. In fact, the data reported after the preliminary analysis was (-110, $+3$, -52, -56)

ns, with mean −54 ns, for the eastward clocks, and (+74, +209, +240, +116) ns, with their mean +160 ns, for the westward clocks (Hafele, 1971). These are the numbers quoted in a defense conference publication in November 1971. However, the numbers were revised to (−74, −55, −57, −51) ns, averaging -59 ± 10 ns, and (284, 266, 277, 266) ns, averaging $+273 \pm 7$, in the published paper in *Science*, submitted in March 1972. The data in the table is from the *Science* paper. This was not consistent with the statement in the publication in November: "The results reported here depend only on the intercomparison data with MEAN(USNO) before and after the trips. The final analysis is not expected to change these results significantly." The MEAN(USNO) refers to the mean time of the high precision clocks at rest in the laboratory. The difference of the analysis in the two papers is striking. However, we will take the final published data in the journal *Science* at the face value, and use that for the discussion.

The observed data from the experiment is summarized in the table. The gravitational time dilation was calculated by the experimenters accurately from the flight data containing altitudes to be $+144 \pm 14$ ns for the eastward clock and $+179 \pm 18$ ns for the westward clock, the 20% difference coming from the 20% difference in flight durations, and partly from the difference in average flight altitudes. The gravitational time dilation expected for the altitude of 9 km is $\Delta T_g/T = gd/c^2 \simeq 3.5$ ns/h. Therefore, even an approximate estimate gives +141 ns and +172 ns as the gravitational time dilation for the eastward and westward trips, close to the more precise values estimated from the flight data. As expected for the gravitational time dilation, these clocks in flights age faster compared to a ground clock, and this is indicated with the positive sign. Once this contribution is subtracted, all the clocks can be compared for purely velocity dependent effects.

The STR time dilation is to be calculated using Einstein's formula

$$\Delta T_v \simeq \int -v^2(t)dt/2c^2 \simeq -T_f\bar{v}^2/2c^2 \qquad (5.19)$$

relative to the clocks in the rest frame of the laboratory. T_f is the flight duration and $\bar{v}$ is the average relative speed. With the recorded average ground speeds and the total duration, this is approximately estimated as 48 ns for the eastward flight and 47

Table 5.2 The observed difference in the duration of the clocks in the flights relative to the stationary clocks in the laboratory, in the Hafele–Keating experiment

	Eastward trip	Westward trip
Total observed (ns) O	-59 ± 10	$+273 \pm 7$
δT_g gravitational (ns) G	$+144 \pm 14$	$+179 \pm 18$
δT_v motional $(O - G)$	-203 ± 17	$+94 \pm 19$
Einstein's prediction δT_v (STR)	-41 ± 8	-49 ± 10
Cosmic Relativity δT_v	-193 ± 8	$+103 \pm 10$

Note: The gravitational time dilation due to the Earth's gravity is subtracted to get the purely motional time dilation. The prediction from the STR as well as the prediction from Cosmic Relativity for the motional time dilation are shown. The minus sign indicates the time dilation (slower time). The positive sign is for the faster ageing of the moved clock relative to the stationary clocks, a phenomenon not known in the STR.

ns for the westward flight. The estimate (by Hafele and Keating) using the actual flight paths is indicated in Table 5.2. All the time comparisons are in nanoseconds, relative to the clock stationary in the laboratory, exactly as done in the experiment. (It is important to note that in the experimental procedure, the clocks are initially synchronized and then finally compared directly after the round trip, with *no intermediate clocks or light signals used for the comparisons or time transfers*.)

The table shows the purely velocity dependent motional time dilations seen in the experiment. After the westward trip, the clocks-ensemble was 94 ± 19 ns *ahead of the stationary clocks* in the laboratory, which corresponds to a *faster ageing of the clocks in motion*. In contrast, after the eastward trip, the clocks-ensemble is 203 ± 17 ns behind the laboratory clocks, indicating time dilation.

The first striking observation is that *the twin who went on a trip has returned older, and not younger as the STR predicts according to Einstein's formula, when he took the westward trip*. On the other hand, when the trip was eastwards, there was excess time dilation, much more than expected from Einstein's formula. The key fact is that the observed motional time dilations match what is predicted by Cosmic Relativity. We have already seen the expression for the motional time dilation of clocks in Cosmic Relativity, in which the cosmic gravitational time dilation depends on the absolute velocity of the motion in the cosmic reference frame. We will now reproduce the simple derivation of the relevant formula, to demonstrate that what was seen in the Hafele–Keating experiment was the cosmic gravitational time dilation of the moving clocks, precisely as described by Cosmic Relativity. I am doing this exercise in full detail, even including aspects that will drop out of the final expression, to point out the exact physical situation.

In Cosmic Relativity, all motional time dilations are gravitational time dilations of clocks moving in the gravitational potential of the Universe. The clock that is stationary in the laboratory at the latitude θ has the velocity relative to the cosmic frame (CF) equal to $\vec{v}_c = \vec{v}_{ss} + \vec{v}_e + \vec{v}_r$, where the first two terms are the velocity of the solar system relative to the CF and the velocity of the Earth in its orbit. The last term is the linear velocity due to the eastward rotation of the Earth, $\vec{v}_r = \vec{r} \times \vec{\Omega} = R_e\Omega\cos(\lambda)$. The time dilation of the laboratory clocks, at rest on the Earth, is

$$\begin{aligned}\delta T/T &\simeq -v_c^2/2c^2 \\ &= -(v_{ss}^2 + v_e^2 + v_r^2 + 2\vec{v}_{ss}\cdot\vec{v}_e + 2\vec{v}_{ss}\cdot\vec{v}_r + 2\vec{v}_e\cdot\vec{v}_r)/2c^2\end{aligned} \tag{5.20}$$

For another clock that is in motion relative to the ground (like an aircraft) at the ground speed $\pm v_f$ (+ sign for the eastward velocity and − sign for the westward velocity), the total velocity in the CF is

$$\vec{v}_c = \vec{v}_{ss} + \vec{v}_e + \vec{v}_r + \vec{v}_f \tag{5.21}$$

and the time dilation in the CF would be

$$\begin{aligned}\delta T_f/T_f &\simeq -v_f^2/2c^2 \\ &= -(v_{ss}^2 + v_e^2 + v_r^2 + v_f^2 + 2\vec{v}_{ss}\cdot\vec{v}_e + 2\vec{v}_{ss}\cdot\vec{v}_r \\ &\quad +2\vec{v}_e\cdot\vec{v}_r + 2\vec{v}_{ss}\cdot\vec{v}_f)/2c^2 \\ &\quad -(2\vec{v}_e\cdot\vec{v}_f + 2\vec{v}_r\cdot\vec{f}_r)/2c^2\end{aligned} \tag{5.22}$$

The reason I write all the terms explicitly is to highlight the fact that all these contributions are factually present in the total time dilations of the two clocks. *When a comparison is made, the common time dilations are not observable, because time can be sensed only as a comparison between the clocks*; then any change that is common to both is not observable, ever. It is a deep fact of physics that there are many real things out there that cannot be observed and measured (like a uniform interaction potential)!

With the common terms dropping off in the comparison of the clocks, the *relative time dilation*, which is the only measurable quantity, is

$$\frac{\delta T_f - \delta T}{T} \equiv \frac{\Delta T_f}{T} = -\frac{v_f^2}{2c^2} - \frac{\vec{v}_{ss} \cdot \vec{v}_f}{c^2} - \frac{\vec{v}_e \cdot \vec{v}_f}{c^2} - \frac{\vec{v}_r \cdot \vec{v}_f}{c^2} \qquad (5.23)$$

The first term involves the relative speed (ground speed) of the flying clocks. That amounts to 1 ns/h, for an average flight speed around 230 m/s. The second term has the large speed of the solar system through the cosmic frame, about 370 km/s. Therefore, its contribution to the time dilation can be as big as ± 3.4 μs/h! The next term with the orbital speed of the Earth, 30 km/s, contributes up to ± 270 ns/h. Each of the last three terms can contribute with positive or negative sign, depending on the relative directions of the velocities involved. Hence, moving clocks can run slower or faster relative to a stationary clock in a reference frame, if the frame itself is moving relative to the cosmic frame. What really matters for the time dilation is the velocity of the clock relative to the cosmic frame.

Then, why don't we see such large loss or gains of duration in the experiment? The answer is found in the most important fact about the comparison of clocks and time; a comparison of clocks is possible only in proximity, at the same spatial location. In the comparison of spatially separated clocks that uses messenger signals between the clocks, the nature of their propagation gets in the way of genuine direct comparison. We have discussed earlier in great detail how the nature of the propagation of light gets intertwined with the comparison of clocks. Alternatively, the clocks can be brought together by transportation. There are only two ways to bring together two clocks that were separated from a common location, for the comparison of the elapsed durations: either reverse their relative motion, as in the two-way experiments, or make them go in a round trip in a looped trajectory, without any reversal of the sense of their motion (Fig. 5.5). In both cases, a term like $\vec{v} \cdot \vec{v}_f$ goes to zero, because v_f takes both positive and negative values relative

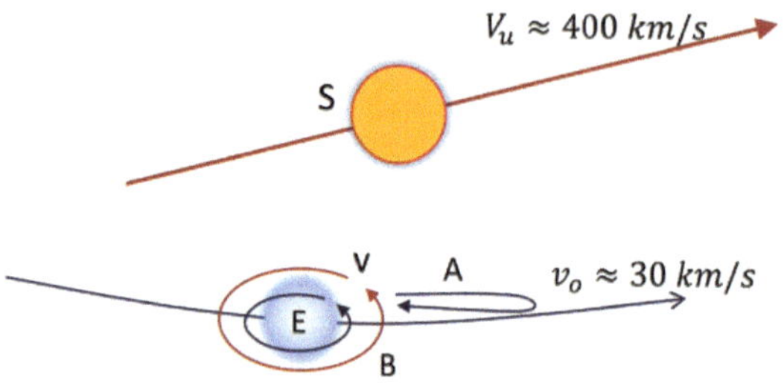

Figure 5.5 The reason why the real time dilations due to the motion of the solar system through the Universe (V_u) and the motion of the Earth in its orbit (v_o) cannot be observed in the terrestrial clock comparison. For the two-way transport (A) of the clocks and for the one-way transport in a round trip around the Earth (B), the velocity of the clocks takes both positive and negative signs relative to the fixed velocities v_o and V_u. Hence, their contribution *averages* to zero in a round trip.

to the global velocity v. Therefore, both the terms, $\vec{v}_{ss} \cdot \vec{v}_f$ and $\vec{v}_e \cdot \vec{v}_f$ average to zero in the round trip, leaving only the first and the last terms in eq. 5.23. In the last term, the speed of the reference clocks is $v_r = \Omega R_e \cos(\lambda)$, where λ is the latitude. The term $\vec{v}_r \cdot \vec{v}_f$ is positive for the eastward motion and negative for the westward motion, and has the magnitude $R_e \Omega \cos(\lambda) v_f$. Thus, we have the full observable motional modification of time in the Cosmic Relativity formula,

$$\begin{aligned} \Delta T_f &= -T\frac{v_f^2}{2c^2} \pm \frac{v_f T R_e \Omega \cos(\lambda)}{c^2} \\ &= -T\frac{v_f^2}{2c^2} \pm \frac{2\pi R_e^2 \Omega \cos^2(\lambda)}{c^2} \end{aligned} \tag{5.24}$$

because $T v_f = 2\pi R_e \cos(\lambda)$, exactly, for the round trip around the globe. This is just the distance travelled in the round trip.

With the parameters of the Hafele–Keating experiment, the first term contributes -49 ± 10 ns for the westward trip. The second term contributes $+152$ ns. Thus, the anomalous faster ageing of the clock due to the cosmic gravitational effect is about 103 ± 10 ns. The observed ageing is $+94 \pm 19$ in striking agreement with this prediction.

Similarly, for the eastward trip, the first term is -41 ± 8 ns, and the second term contributes with the same sign, resulting in the time dilation of -193 ± 8 ns, to be compared with the observed -203 ± 17 ns.

Thus, there is no doubt that the motional time dilation observed in the Hafele–Keating experiment is the cosmic gravitational time dilation of the clocks moving relative to the vast amount of matter in the Universe.

The striking and significant result that the *westward clocks in fact lost time, and ran faster than the immobile clocks in the laboratory by about 94 ns,* should have been an eye opener and a strong clue for the reconsideration of the entire basis for relativistic physics. But, alas, that did not happen. The twin had returned older after the round trip, contradicting the STR directly. Yet, the results were celebrated as the 'proof of the time dilation in Einstein's theory'. The experimenters tried to justify the result by accusing the Earth of being a bad non-inertial frame. The Earth was apparently not inertial enough to apply the Special Theory of Relativity! Therefore, they tried to argue that the calculation using just the relative velocity of the flights, relative to the laboratory in which the comparisons of the clocks were done, was not sufficient. Then they decided that the data were to be analysed from a frame in which the motion of the laboratory due to the rotation of the Earth was explicit. For this, an Earth-centred non-rotating frame was chosen. (Notably, nobody was bothered about the non-inertial nature of the orbital motion of the Earth.) Thus, in effect, the experimenters used the expression from a theory of relativity with an absolute frame to calculate the time dilation, which of course agreed with the observed time dilation. The general impression today is that the Hafele–Keating experiment with the atomic clocks somehow verified Einstein's predictions, not only for the gravitational time dilation but also for the motional time dilation of the STR.

However, now we have to be alert, because there is an obvious inconsistency in such a view, apart from the factual discrepancy that is evident (see Table 5.2). Suppose we grant that the measured time dilations are according to the prevailing theory of Einstein's. But, now that the new knowledge of *the matter in the Universe at nearly the critical density is established from observations,* surely there is a contribution from its gravity as well. We can and should estimate it, at least approximately. As we already did, that also is comparable to what Hafele and Keating calculated from the 'non-rotating earth-frame', using Einstein's theory in empty space. Then, we face the indigestible fact that we cannot have both the Einstein time dilation as well as the cosmic gravitational time dilation. *That would double the expected modifications of time, which will contradict the experimental results.* We have to definitely choose one of these theories. However, the matter in the Universe is observed, and its average density is measured by different means. We cannot deny the existence of the observed matter and the fact that it has gravity! Then, the choice of the correct theory is obvious. The honourable opportunity to realize that the STR was not the correct theory was available 50 years ago, but in the absence of the gravitational paradigm, no alternative was in sight. The situation has changed with Cosmic Relativity and any hesitation on the right choice is least justified.

5.9 Summary

We have gained a sense of satisfaction, of understanding completely how and why exactly the time marked by the clocks in motion differ from that of the clocks at rest. First, we convinced ourselves about the qualitative and quantitative aspects of the enormous gravity of the vast amount of matter in the Universe. Then we found that the gravitational potential experienced by a clock at rest and a clock in motion are different, by the factor $\gamma = 1/\sqrt{1 - v^2/c^2}$. Since we already knew about gravitational time dilation, it was an easy step to realize that all motional time dilations are really gravitational time dilations due to the matter in the Universe. *That was a quantum leap in our understanding time and its modification.* We will not miss the demise and loss of the old theory, which we held dearly for over a century. That is insignificant when compared to the level of enlightenment, and the precision achieved in the description of motional time dilation.

I discussed the Hafele–Keating experiment, as a lost opportunity of the past to correct the course of the physics of relativity. The fact that clocks in motion could age faster, and not slower as the STR predicted and insisted, should have been an eye-opener, but it turned out to be a misinterpreted reason for closing further scrutiny, unfortunately.

Now we are taking an interlude. We want to be sure of our path in this exciting journey. After all, I am making the extraordinary statement that the Einstein paradigm of the STR needs to be entirely replaced by my cosmic gravitational paradigm of Cosmic Relativity. *There is no better way to be absolutely sure than to rely on experimental results.* I promised to show the firm experimental results that the relative speed of light is Galilean, and not an invariant constant. Such a direct proof should remove any lingering doubt. So, in this interlude, I will present the results of a novel experiment that studied the nature of the genuine one-way propagation of light, and found that it is indeed Galilean. However, the presentation of my results will be strengthened considerably if we examine the writings of Einstein on simultaneity in his theories of relativity, because the comparison of clocks and the synchronization of clocks using light depend on the exact nature of the propagation of light. Then we will realize, to our disbelief, that *Einstein himself had made a crucial error in the discussion of simultaneity* that made his Special Theory of Relativity inconsistent! Nobody would be ready to believe that possibility, but the facts are bare and easy to examine, and verify. Besides, the error was pointed out with clarity to Einstein in a direct confrontation earlier, by none other than Henri Bergson, the well-known philosopher who studied time, duration, and simultaneity in Einstein's theory, with remarkable logical and mathematical integrity (Bergson, 1922).

Chapter 6

Interlude on Simultaneity and the One-Way Speed of Light

The concept of simultaneity is of central importance in the theories of relativity, being an essential prerequisite for defining the synchronization of clocks and the measurement of time. As emphasized by Einstein, the time of an event can be reliably attributed only by consulting a 'local' clock, adjacent to the event. Then, one has to specify a method to synchronize such clocks at different locations, to build a consistent physical theory. That was how the constancy of the velocity of light in all inertial frames became the pivotal point of relativity. After postulating the invariance of the velocity of light, Einstein asserted that if all clocks are synchronized using light, in the rest frame of the clocks, they are all properly synchronized, independent of any common uniform velocity they may have. In other words, they are synchronized as if they are all at rest.

In effect, Einstein's light hypothesis, of the constancy of the relative velocity of light, took out the variable from the physics of relativity. If the hypothesis is factually true, space and time are the same, both conceptually and physically, with only a constant conversion factor between them. Otherwise, space and time are two

Gravity's Time
C. S. Unnikrishnan

ISBN 978-981-4877-72-5 (Hardcover), 978-1-003-25656-4 (eBook)
www.jennystanford.com

separate and independent entities. This is of utmost importance when one formulates the various measures of physical quantities and their units. For centuries and millenniums, we had 'length' and 'time' as separate fundamental quantities, with separate methods, gadgets and units for their measurements. With Einstein's STR, they were fused into one, with the invariant speed of light 'c' as the conversion factor. The international convention of measures and units adopted this in defining the 'standards' of measurements. The facilities like the GPS were set up on this premise. One can see that the light hypothesis that defines simultaneity, and the process of clock-synchronization based on the hypothesis, are all-pervading. Therefore, even the slightest falsity of the light hypothesis would be fatal for relativity and wider physics.

We have already seen in multiple contexts that all the empirical evidence hitherto is consistent with both the Galilean propagation of light as well as the light hypothesis, because the two-way experiments as well as the one-way experiments with spatially separated clocks on the propagation of light cannot distinguish between the two antipodal scenarios. Given this fact, what exactly is the physical truth?

We will find the definite answer now.

6.1 Einstein's Train and Platform Scenario

Any discussion of the study of the relative velocity of light needs to understand the notion of simultaneity, which is relevant for the synchronization of clocks. We will start with Einstein's own discussion about simultaneity and its physical criterion in his theory. Einstein had discussed and clarified the notion of simultaneity in the STR in his popular monograph *Relativity: The Special and General Theory* (Einstein, 1920). He used the example of a railway carriage (train) and an embankment (platform) on many occasions in the discussions on relativity. In the chapter 'The idea of time in physics', he wrote, 'Lightning has struck the rails on our railway embankment at two places A and B far distant from each other'. He clarified that by 'time of an event', one means the reading of a clock which is in the immediate vicinity (in space) of the event. Thus, a time-value could be associated with every event. What remains is the issue of synchronization of clocks at different positions. Then, one can judge whether two distant events are simultaneous or not, relative to such clocks and the synchronized time.

Of course, events that are occurring in distant locations can be sensed only with the help of messenger waves or material bodies that propagate between the locations of the events and the observer. This is why the nature of their propagation gets intertwined with the synchronization of time. Sound is a familiar messenger of events, with great importance for all life forms. But, light is the most common and effective messenger wave, and it is also the fastest. With the miraculous property of propagation at the constant relative velocity that Einstein conferred on light, it becomes a unique messenger of distant events. Then, Einstein discussed a core question of relativity: if two events are simultaneous for one observer (reference frame), will they be simultaneous also for another who is moving relatively? He wrote,

> We suppose a very long train travelling along the rails with the constant velocity v. . . . Then every event which takes place along the line also takes place at a particular point of the train. . . . As a natural consequence, however, the following question arises:
>
> Are two events (e.g. the two strokes of lightning A and B) which are simultaneous with reference to the railway embankment also simultaneous relatively to the train?

The problem is clearly stated (Fig. 6.1). We will answer this in steps, first assuming the Galilean propagation of waves, like sound (or light in Cosmic Relativity). Consider two observers O and O' in

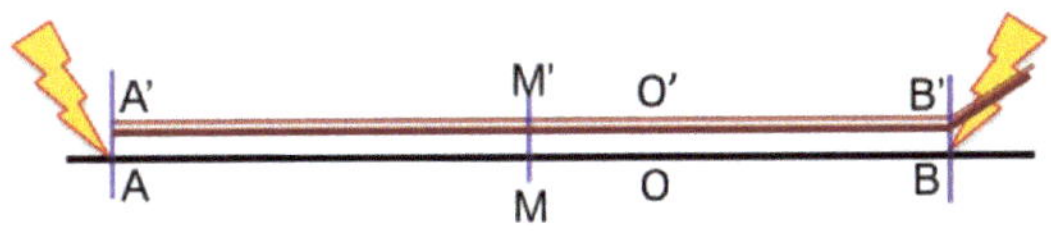

Figure 6.1 Einstein's depiction of the reference systems for discussing the relativity of simultaneity. The train A'B' and the railway platform AB are in relative motion, with 'midpoint' observers O' at M' and O at M (adapted from the figure in Einstein's monograph (Einstein, 1920), with added labels for better clarity).

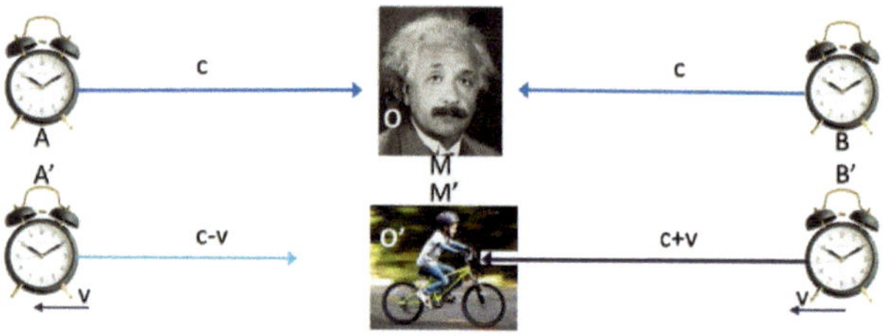

Figure 6.2 The situation regarding the apparent simultaneity with Galilean waves (sound) as the messenger of events that are spatially separated. A pair of distant events that are *simultaneous* to a stationary observer O become *successive* relative to an observer O' moving at the velocity v. In the frame of O', the relative velocity of sound is $c + v$ from B' and $c - v$ from A'. Also the sources appear to move relatively.

relative motion. O is stationary at the midpoint M relative to the two events and the sources of the Galilean waves at A and B, say, the ringing from two alarm bells that are sounded at the same time (Fig. 6.2). Therefore, O hears the bells simultaneously, say at the time $t = T_0$. Conversely, if O who is stationary at the midpoint hears the bells simultaneously, then it is deduced that they were sounded at the same time ($t = T_0 - d/c_s$) at A and B. The delay or 'retardation' d/c_s due to the finite speed of the waves can be corrected for, though this is not required to judge whether the events are simultaneous or not.

The observer O', who coincides with O at M'=M when the bells were sounded, is in motion towards B at the relative speed v, *as judged from O*. However, in the rest frame of O', it is the observer O and the alarm bells that are relatively moving. More importantly, the sound waves are arriving at different relative speeds $c_s + v$ from B' and $c_s - v$ from A', the source points in the frame O'. Therefore, O' will hear the bell from B' first, at $t'_1 = L/(c_s + v)$, and that from A' later, at $t'_2 = L/(c_s - v)$; then *the events would be judged as not simultaneous.* I remind the important fact that the motion of the source points, at distance L when the signals were emitted, does not affect the relative velocity of the sound waves, just as for light. The relative velocity of sound (and of waves in water, etc.) depends only on the velocity of the observer (relative to the medium) and not on the velocity of the source (I state this explicitly because it is a much misunderstood fact). Thus, the unambiguous answer to Einstein's query for the case of Galilean waves is that different observers perceive the order of events in time differently, depending on their state of motion. *Simultaneity, defined through the arrival of the Galilean messenger waves, is relative and not universal.* For the Galilean messenger waves with velocity c relative to a stationary reference, the lack of simultaneity of events separated by the distance $2L'$ is

$$\Delta t' = \frac{L'}{c-v} - \frac{L'}{c+v} = \frac{2L'v}{c^2(1-v^2/c^2)} \tag{6.1}$$

For $v \ll c_s$ the second-order term in brackets can be ignored. *Events that the stationary observer perceives as simultaneous will not be simultaneous for the observer in motion, and vice versa.* An identical conclusion is reached from the rest frame of the stationary person as well; the moving person moves away from the midpoint and therefore the sounds of bells reach him in succession from the locations B and A. There is consistency and there is no reciprocity.

We note an important additional point that is obvious in the case of sound. Not only that the observer O' perceives the bells in succession, he would also perceive them shifted relatively in pitch; the one from B' at a higher pitch than the one from A', due to the Doppler effect. Hence, he has a measure of his real motion and velocity relative to air. Therefore, absolute simultaneity is recoverable from the full physical data when the messenger of the spatially separated events is Galilean.

Now we analyse Einstein's query on simultaneity with light as the signal, *under the assumption that the relative velocity of light is the same in all inertial frames, independent of the velocity of the frames,* as Einstein had postulated in the STR. Due to this cardinal difference from the familiar Galilean waves like sound, we should naturally expect a different conclusion regarding simultaneity, when light is the messenger wave, instead of sound. In other words, if the relative velocity of sound were the same invariant constant for the stationary as well as the moving observer in the example of the alarm clocks, we would have reached a different conclusion.

The observer O, stationary at the midpoints of the events, receives the signals at the same speed c from the equal distances L. Therefore, the events are perceived as simultaneous. The observer

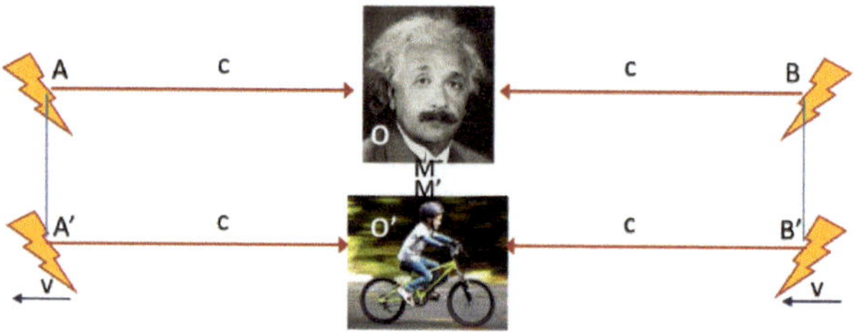

Figure 6.3 The universal and invariant speed of the messenger waves (light) relative to any inertial observer in the STR implies universal simultaneity. The relative velocity of light does not depend on the relative velocity of the source. So, *in the rest frame of an observer*, if the waves were emitted when he is at the midpoint of the events, the waves will reach simultaneously, whether the sources are moving or not! Therefore, in the STR, in which any inertial observer can claim the state of rest, simultaneity is universal. But, this is inconsistent with the deduction of the observer O about what O' perceives, and vice versa.

O' also receives the light signals at the same relative speed c. In his rest frame, the sources (as well as the observer O) are moving. But, *the relative speed of light does not depend on either the motion of the sources or the observer, according to the STR*. Therefore, the light from those moving sources, which were equidistant when the signals were emitted, will be received simultaneously by the observer O' as well (Fig. 6.3). Two events that are simultaneous to O are simultaneous to O' as well, when the relative velocity of light is a universal constant, independent of the motion of the observer. This should not be surprising because every inertial observer in the STR has the equal right to claim the state of rest.

Coming to Einstein's specific scenario, in the frame O' of the train (Fig. 6.1), the observer at the midpoint M' is at rest, in his frame. After lightning strikes A' and B', light travels at equal speed from the equidistant points A' and B' towards M' (the relative velocity of the source does not affect the relative velocity of light). The light from A' reaches at time $t_1' = L'/c$ and the light from B' reaches at $t_2' = L'/c$. Evidently $t_1' = t_2'$ and $\Delta t' = 0$. Then, the light flashes arrive *simultaneously at M'* and the *observer perceives the events as simultaneous*. Now, we consider the observer who is at rest in the frame O, on the platform. Relative to O, the train O' is moving, but that is irrelevant for what the observer experiences in the frame O. Light pulses travel at equal speed from the equidistant points A and B towards the midpoint M. $t_1 = t_2 = L/c$. They will reach *simultaneously at the observer at M*, who perceives the lightning at A and B as simultaneous. So, we conclude that the *experiences of observers in O and O', in their own rest frames, are identical, assuming the universal relative velocity of light.*

However, this raises a serious problem. For the observer O', the other observer O is moving away from the midpoint of the simultaneous events. But, for the observer O, it is O' who is moving away from the midpoint. So, *each observer would reckon that the other would receive the light flashes at a point that is away from the midpoint of the events*, and hence in succession, and not simultaneously! Therefore, there is no consistency between the physical experience and the simple theoretical (geometrical) deduction, in a theory in which the relative velocity of light is an invariant constant.

We note in contrast that if light propagated similar to sound and if its relative speed was *not a universal constant*, the events perceived as simultaneous in one frame would have been seen as one after the other, in succession, in another frame, relatively moving. The lack of simultaneity in the moving frame is given by eq. 6.1, and in the 'stationary frame' by $\Delta t = Lv/c^2$. This is the situation in a theory of relativity with a master (absolute) frame. There is complete mutual consistency.

What was Einstein's answer, which is also today's physicists' answer and belief? Continuing with what Einstein wrote,

> When we say that the lightning strokes A and B are simultaneous with respect to the embankment, we mean: the rays of light emitted at the places A and B, where the lightning occurs, meet each other at the mid-point M of the length A–B of the embankment. But the events A and B also correspond to positions A and B on the train. Let M' be the mid-point of the distance A–B on the travelling train. . . . *Now in reality (considered with reference to the railway embankment) he is hastening towards the beam of light coming from B, whilst he is riding on ahead of the beam of light coming from A. Hence the observer will see the beam of light emitted*

> *from B earlier than he will see that emitted from A.* Observers who take the railway train as their reference-body must therefore come to the conclusion that the lightning flash B took place earlier than the lightning flash A. We thus arrive at the important result:
>
> *Events which are simultaneous with reference to the embankment are not simultaneous with respect to the train, and vice versa (relativity of simultaneity).*

Einstein's conclusion was identical to the one we arrived at *in the case of the Galilean waves* (like sound), though he assumed the invariant relative speed of light! Einstein had made a crucial error of judgment while formulating the central theme of simultaneity and clock synchronization in relative motion. He did not realize that his postulate of the invariant relative velocity of light was not consistent with his assertion on simultaneity. I have italicized some sentences to emphasize the faulty logic in Einstein's analysis. The reading of these passages by generations have not spotted this logical misstep and why the conclusion with light (of invariant velocity) ended up in preferential simultaneity for the observer on the embankment, and events in succession for the observer in the train, in a theory in which both observers are equivalent. They have the same speed of light, and either one can consider himself as at rest and the other one as moving. The fatal gap of logic in Einstein's analysis is easily spotted with some guidance. There is no 'real motion' in the STR; it is a theory in which only relative motion makes sense. Einstein errs within his own theory when he writes 'in reality he is hastening towards the beam of light coming from B'. *The observer is at rest in his frame and the source of light is coming towards him, but that does not affect the relative speed of light, or of any wave.* If Einstein had started his discussion with the train as the reference, he would have reached the conclusion that the events were simultaneous to the observer in the train, but not to the observer O on the embankment. Then the train is the rest frame and the embankment is moving relatively. O' would receive the signals simultaneously, being at rest at the midpoint M' of the events (Fig. 6.1). But, O would receive them in succession, because O moved away from M' and hastened towards point A', relative to the immobile train! It is clear that Einstein would have been saved from this fatal error of inadvertent preference for the embankment as 'really stationary' if he had considered two trains in relative motion, instead of a train and an embankment. For sound, there is the medium of air that defines who is a really stationary observer. But, for light in the STR, there is no such reference to define a state of absolute rest. By spotting this error, we can see that Einstein's light hypothesis is in conflict with the notion of simultaneity and clock synchronization in the STR. Einstein's Special Theory of Relativity is clearly inconsistent, as per the testimony of its creator himself (Unnikrishnan, 2018a).

What Einstein describes is not what the observer in the train experiences, but what the person on the platform, who considers M' as moving, confers on the moving observer as his experience. However, he failed to notice that this assumed experience is inconsistent with the light hypothesis. On the one hand, the experience of simultaneity for both observers should be identical if the light hypothesis is used. But on the other hand, each observer is moving relative to the other towards one of the sources, and hence should receive that signal earlier. However, the same pair of signals cannot be both simultaneous and successive. This proves that the light hypothesis is inconsistent.

If all this seems unbelievable, recasting the scenario in terms of observers in relative motion in two trains, whose midpoints coincide at the moment when the two lightning events take place, should clarify (Fig. 6.4). Take the observer O' first, for a change. He is at rest in his frame, and O is moving relatively. Since he was at the midpoint when the events happened, the source A' moves away from the midpoint M', and B' moves towards it. But that does not affect the speed of the signals of the event that happened already! So, propagating at equal speeds they arrive simultaneously at M'. Exactly the same argument applies from the rest frame of O, in the other train. This symmetric example bares Einstein's simultaneity fallacy that invalidates his Special Theory of Relativity beyond any doubt.

The philosopher Henri Bergson had found the fallacy in Einstein's logic, while trying to reconcile the common sense notion of simultaneity and the role of the concept in the physical theory, and expressed his disagreement when they met in a discussion session in the Sorbonne in Paris, in April 1922 (Nordmann, 1922). However,

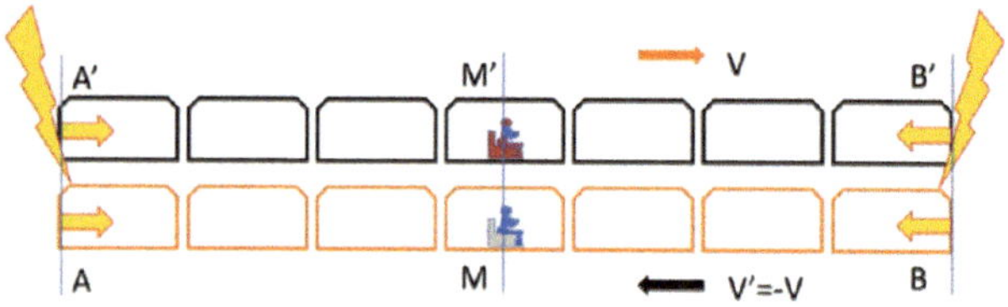

Figure 6.4 Observers in the two trains coincide at M and M' at the moment of the events at A = A' and B = B'. Trains are in relative motion at the relative velocity v. There is no difference between the two, and there is no preference for the observer at rest. This scenario in the STR is entirely symmetrical. Then, there is no difference in the claim for simultaneity either.

the brief remarks by Bergson were not seriously considered by Einstein.

Bergson wrote in his monograph, *Duration and Simultaneity* (Bergson, 1922), published soon after he had expressed his reservations about Einstein's theory, citing Einstein's example,

> This passage enables us to catch on the wing an ambiguity that has been the cause of a good many misunderstandings... we must not forget that the train and the track are in a state of reciprocal motion... Let us now emit our two flashes of lightning. The points from which they set out no more belong to the ground than to the train; the waves advance independently of the motion of their source.
>
> It then becomes evident at once that the two systems are interchangeable, and that exactly the same thing will occur at M' as at the corresponding point M. If M is at the middle of AB and if it is at M we perceive a simultaneity on the track, it is at M', the middle of B'A', we shall perceive this same simultaneity in the train... what is simultaneity with respect to the track is simultaneity with respect to the train.

Bergson was impeccable in his logic concerning simultaneity in Einstein's theory (Unnikrishnan, 2020b). Bergson showed that the asymmetrical conclusion on simultaneity that Einstein arrived at was merely a result of the preferential choice of the frame of the platform as 'stationary' and that of the train as 'moving'. If we had started with the train as the rest frame and the platform in relative motion, we would come to the opposite conclusion. Since both the frames are equivalent, their experiences of simultaneity are identical.

Since it is straightforward and elementary to analyse the issue of simultaneity of distant events, because only the first-order (v/c) effect of the propagation of light is involved, everybody who is alert and patient can arrive at the correct conclusion by progressing step by step through the arguments, paying attention to the assumptions. Einstein's own writings and his example have led us to the internal inconsistency of his Special Theory of Relativity.

Why didn't anybody, other than the lone figure of Henri Bergson, insist on logic and rigour in analysing this central issue of simultaneity in relativity? I have no explanation, except to say that the discussion at the Sorbonne was too brief for anybody to understand the exact point raised by Bergson, and later, nobody who mattered seemed to have examined the contents of Bergson's remarkable monograph, *Duration and Simultaneity*, for the details. In fact, when I saw the dismissive comments about Bergson's stand in a recent book on time and relativity, I asked one of the authors whether they had read Bergson's arguments in *Duration and Simultaneity*. The answer was an emphatic 'No'!

6.2 The Relative Velocity of Light

6.2.1 General Considerations

We discussed in detail the fact that experiments in the category of the Michelson–Morley experiment and its variations are not sensitive to the nature of propagation of light (Section 3.2). These two-way experiments, where light propagates along and against the motion of the apparatus, are consistent with both the Galilean propagation (relative velocity $c \pm v$) as well as the propagation at a constant relative velocity, postulated in the STR. Hence, *one needs to think of an experiment that studies the 'one-way propagation'*. However, such an experiment requires two clocks that are separated in space. Then the subtle synchronization issues that we analysed come to the fore (Section 3.7). The conclusion was that neither the M-M type experiments, nor any of the 'one-way experiments' have managed to reveal the exact nature of the propagation of light.

The physical reason underlying the inability of the two-way experiments to determine the nature of propagation of light is clear: such experiments have only a second-order (v^2/c^2) additional delay in the propagation, if the relative velocity depends linearly on the velocity of the observer. Then, other second-order physical effects like the length contraction or time dilation can mask the true nature of propagation. However, a first-order signal, proportional to v/c, cannot be obscured by a second-order physical effects. But, only an experiment dealing with the one-way propagation can have a first-order signal.

It might be worth repeating the proof of the *impossibility of testing the constancy of the one-way speed of light by schemes that use two spatially separated clocks*. The measurement needs fixing the spatial distance L between two points A and B, and measuring the duration of propagation ΔT of light from A to B, using synchronized clocks. Without synchronization, the physical 'duration' remains arbitrary. Fixing L with a two-way measurement is easy and does not interfere with the measurement of the relative one-way speed, first order in v/c, because any relativistic correction to the length due to the unknowable inertial speed of the reference frame is second order in v/c. But, how do we synchronize the two clocks? Obviously, we cannot use a pulse of light from one point to another to synchronize clocks because the duration depends on what we assume for the propagation of light; if we assume the constant relative speed, the duration of propagation is L/c, whereas if we assume the Galilean propagation, we get the duration $L/(c \pm v) \simeq (L/c) \mp (L/c)(v/c)$, with the first-order correction. Since we cannot know the velocity of the frame (principle of relativity), our synchronization depends on our assumption and chosen convention. Our task of finding out the nature of propagation of light would become logically circular if we used light to synchronize the clocks.

The other possibility of synchronizing two clocks at the same location A and transporting one slowly to point B does not work either. For a clock transported at the small velocity v_c in the frame moving at the velocity v_f, the second-order time dilation is

$$\Delta T_B \simeq -T_B\frac{(v_f+v_c)^2}{c^2} \simeq -T_B\left(\frac{v_f^2}{2c^2}+\frac{v_c^2}{2c^2}+\frac{v_f v_c}{c^2}\right)$$
$$= -T_B\frac{v_f^2+v_c^2}{2c^2} - \frac{Lv_f}{c^2} \quad (6.2)$$

Therefore, there is an *unknowable* first-order time dilation between the two clocks, of $(L/c)(v_f/c)$, making the clock B to run ahead of A, if synchronized with this protocol. If we use such a clock to measure the propagation duration of light, the synchronization mismatch $(L/c)(v_f/c)$ masks the Galilean propagation factor $-(L/c)(v_f/c)$, giving the false conclusion that the speed of light is an invariant! Therefore, *it is impossible to decide whether the relative speed of light is Galilean, or an invariant constant, in a one-way measurement that employs two spatially separated clocks.*

Even seasoned experts do not always realize this crucial point of synchronization gap that we repeatedly discussed. One can find many prominent publications about experiments that studied the one-way propagation of light, with the conclusion that they verified the isotropic nature of propagation of light in the STR (its invariant relative velocity). The irony and embarrassment of the situation go deeper than one might imagine. A brief glance at the relevant publications will clear the air.

The need for one-way measurements to demarcate Einstein's theory from the Lorentz–Poincaré theory was reiterated in a paper in the journal *Physical Review Letters* by M. Ruderfer in 1960 (Ruderfer, 1960). He suggested a one-way experiment with gamma rays, on a rotating platform, based on the newly discovered Mössbauer effect; a definite velocity-dependent signal would favour the Lorentz–Poincaré theory and a null signal (zero) would support Einstein's theory. Very soon Ruderfer published a vital erratum (Ruderfer, 1961), when he understood the fundamental issue of the inseparability of the propagation delay and the duration measured with separated clocks. The erratum clarified that both the LPTR and the STR predicted identical null signal, making such one-way experiments ineffective. This fundamental 'catch' was already known and discussed by Poincaré, stressing the role of 'convention'

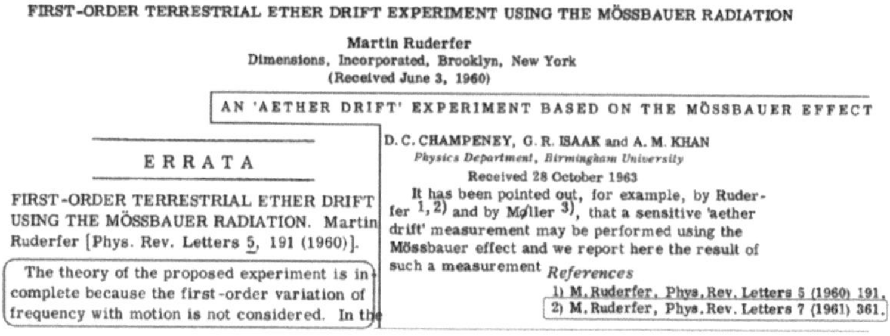

FIRST-ORDER TERRESTRIAL ETHER DRIFT EXPERIMENT USING THE MÖSSBAUER RADIATION

Martin Ruderfer
Dimensions, Incorporated, Brooklyn, New York
(Received June 3, 1960)

AN 'AETHER DRIFT' EXPERIMENT BASED ON THE MÖSSBAUER EFFECT

ERRATA

FIRST-ORDER TERRESTRIAL ETHER DRIFT USING THE MÖSSBAUER RADIATION. Martin Ruderfer [Phys. Rev. Letters 5, 191 (1960)].

The theory of the proposed experiment is incomplete because the first-order variation of frequency with motion is not considered. In the

D. C. CHAMPENEY, G. R. ISAAK and A. M. KHAN
Physics Department, Birmingham University
Received 28 October 1963

It has been pointed out, for example, by Ruderfer [1,2] and by Møller [3], that a sensitive 'aether drift' measurement may be performed using the Mössbauer effect and we report here the result of such a measurement

References

1) M. Ruderfer, Phys. Rev. Letters 5 (1960) 191.
2) M. Ruderfer, Phys. Rev. Letters 7 (1961) 361.

Figure 6.5 The brief history of the mishap with the experiments on the one-way propagation of light. Only one experiment is cited, but all subsequent experiments have the same basis, mentioned in M. Ruderfer's original proposal (top). The experimenters obviously were not aware that their second citation (right bottom) was an erratum published by Ruderfer that nullified the proposal and their experiment!

in synchronization, but it is a subtle and deep issue that is easily missed, as history shows (Fig. 6.5). Oblivious to the content of the erratum, researchers went ahead and 'confirmed the light hypothesis', while citing both Ruderfer's proposal paper as well as the nullifying erratum (Champeney *et al.*, 1963)! And many are continuing in vain to refine these tests on the 'isotropy of the speed of light' (Krisher, 1980), still remaining in the dark about the inefficacy of such experiments due to the interdependence of synchronization and time dilation of separated clocks in moving frames.

6.2.2 One-Way Experiments in Rotating Frames

In 1913, Georges Sagnac used a closed-loop optical interferometer at the Sorbonne in Paris to measure the effect of rotation through the 'stationary ether'. He presented the positive results, 'the proof of the reality of the luminiferous aether by the experiment with a rotating interferometer', in several publications (Sagnac, 1913a,b). He considered his result as the decisive disproof of the etherless Special Theory of Relativity (Unnikrishnan, 2020b).

A. A. Michelson had published a similar idea in 1904, to detect the Earth's rotational motion relative to the ether, but managed to do an experiment only in 1924, with H. G. Gale (Michelson, 1904; Michelson and Gale, 1925). Michelson concluded from their

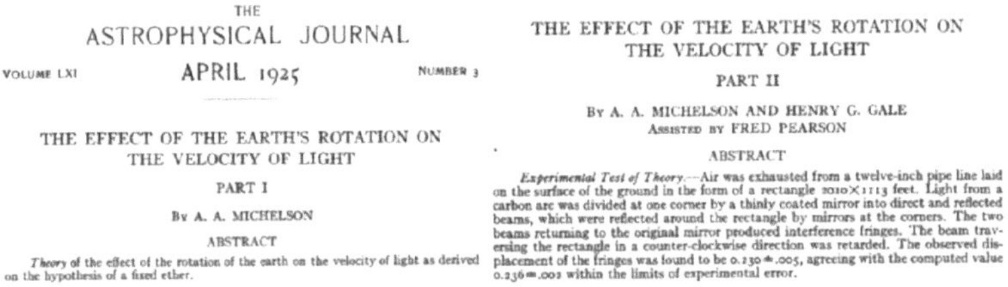
THE
ASTROPHYSICAL JOURNAL
VOLUME LXI APRIL 1925 NUMBER 3

THE EFFECT OF THE EARTH'S ROTATION ON THE VELOCITY OF LIGHT

PART I

BY A. A. MICHELSON

ABSTRACT

Theory of the effect of the rotation of the earth on the velocity of light as derived on the hypothesis of a fixed ether.

THE EFFECT OF THE EARTH'S ROTATION ON THE VELOCITY OF LIGHT

PART II

BY A. A. MICHELSON AND HENRY G. GALE
ASSISTED BY FRED PEARSON

ABSTRACT

Experimental Test of Theory.—Air was exhausted from a twelve-inch pipe line laid on the surface of the ground in the form of a rectangle 2010×1113 feet. Light from a carbon arc was divided at one corner by a thinly coated mirror into direct and reflected beams, which were reflected around the rectangle by mirrors at the corners. The two beams returning to the original mirror produced interference fringes. The beam traversing the rectangle in a counter-clockwise direction was retarded. The observed displacement of the fringes was found to be 0.230±.005, agreeing with the computed value 0.236±.002 within the limits of experimental error.

Figure 6.6 Michelson's paper on the 'theory of the effect of the rotation of the Earth on the velocity of light, as derived on the hypothesis of a fixed ether', and the announcement of the verification of the theory in the Michelson–Gale experiment.

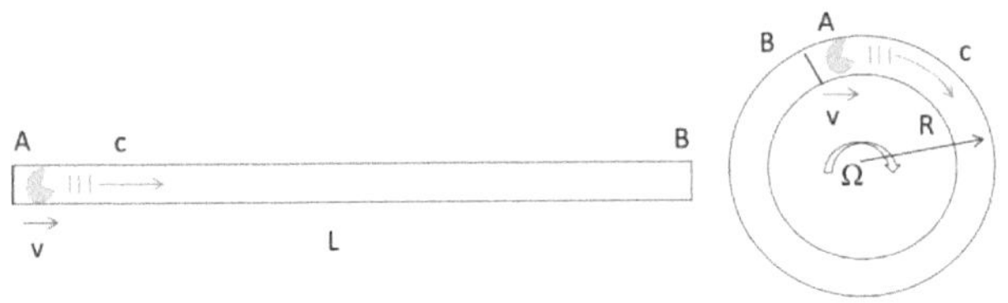

Figure 6.7 The equivalence of the relative durations the wave takes to propagate through a relative distance L, while the frame is moving at the uniform speed v, for the linear case (left) and for the looped circular case (right). Here, $2\pi R = L$, and $R\Omega = v$. The experiment on the left needs synchronizing two spatially separated clocks. The one on the right is the central idea used by Sagnac 1913, as well as by Michelson (and Gale) in 1925, to announce the 'variation of the relative velocity of light due to the motion relative to the ether' (absolute frame).

heroic experiment, with a partially evacuated kilometre-size closed rectangular interferometer, that *the slow rotation of the Earth relative to the stationary ether was detected* (Fig. 6.7).

6.3 Experimental Proof of the Galilean Relative Velocity of Light

I will now discuss the central experimental result from my laboratory at the Tata Institute of Fundamental Research (TIFR, Mumbai) that proves the Galilean nature of the relative speed of light, relative to inertial frames (Unnikrishnan, 2009, 2011). To establish a relative velocity, we need to measure the duration for propagation from point A to B with a spatial distance L measured along the path. As we clarified already, it is immaterial whether there is a length contraction or not when the frame moves at different velocities, because the effect is second order in velocity. Since we are trying to assert whether the relative one-way velocity is a constant c or the variable $c(1 \pm v/c)$, an effect of order v^2/c^2 is insignificant in the experiment (in terms of numbers, $v/c \sim 10^{-8}$, whereas $v^2/c^2 \sim 10^{-16}$).

Once the path length is fixed, the measurement involves only the duration $T(v)$ of propagation between A and B of fixed path length, as a function of the velocity v of the frame, which is the common synchronous velocity of A and B, or the emitter and the detector (Fig. 6.8). The relative speed is $c(v) = L/T(v)$. We measure $T(v)$ either directly or interferometrically. With a fixed oscillator frequency, and no relative motion between the emitter and detector (no Doppler shift), the measurement of the wave-phase is the same as the measurement of time (duration) ($\delta T = \omega \delta \varphi$).

As stressed earlier, we need to adhere to two conditions: (a) the clock synchronization problem needs to be solved, (b) the reference frame should be in inertial motion. I succeeded in solving both issues, resulting in a decisive experiment that falsifies Einstein's light hypothesis.

We note that *light takes identical durations to travel identical path lengths irrespective of the shape of the path,* relative to the same frame. That is, for the one-dimensional path with a bend, L-shaped, the time of propagation is the same as that for the straight path of the same length for the same inertial speeds of A and B, but for a possible small fixed bend-correction that is independent of the velocity of the observer (Fig. 6.8). A special one-way path with a finite number of bends is the closed 'stadium' path with long inertial sections, over which the source and detector move, always inertially. The distance between the emitter and the pulse of light increases, whereas the distance to the receiver decreases monotonically, as in a linear path. So, *we have established logically that the experimental results on the relative speed would be identical in the straight path and in the stadium path.* We shall verify this assertion experimentally as well, to avoid any residual misunderstanding.

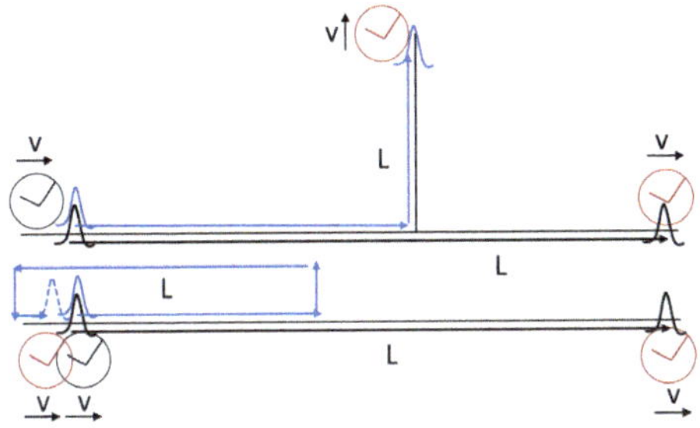

Figure 6.8 Upper panel: The equivalence of the duration of light propagation on a straight path and another path with bends, between two clocks with fixed separation *L*. Lower panel: When the path with bends completes a loop, the clocks with the separation *L* for the propagation of the signal are in fact spatially nearby, solving the synchronization problem. Then, the true relative duration for the one-way propagation of light along the inertial section of the looped path is measurable.

By establishing the equivalence of the results in the linear path with separated clocks, and in the 'stadium path' with the clocks at the 'start' and the 'finish' points, I have removed the crucial hurdle that prevented hitherto the direct measurement of the true one-way relative speed of light, relative to inertial observers. Note that in the 'stadium' configuration, we have a fixed separation *L* between the clocks and yet they are nearby due to the looping, allowing unambiguous synchronization that is free of conventions or postulates. Also, because they are nearby, it is easy to move them together *inertially* on the same short platform. In fact, the two clocks can be mathematically and physically identified and replaced by a single reference clock. *Waves travel the same fixed distance L from the source to the detector in a round trip, without retracing back their path* (Fig. 6.8). *We have indeed solved the crucial problem of clock synchronization, while preserving inertial motion* (Unnikrishnan, 2011).

Relative to the moving frame, the path length is fixed as *L*. This is evident since the source and the detector that define the path of length *L* between them, are then in the 'rest frame' and there is no relative motion between them. The duration depends on whether the relative velocity of light is Galilean or an invariant. If the relative velocity is an invariant (same c_0 in all inertial frames) then δT =

L/c_0. On the other hand, if the relative velocity is Galilean $c_0 - v$, then the duration $\delta T'$ is *the same as in the lab-frame to first order in* v/c. (For Galilean physics $\delta T' = \delta T$ to first order in v/c).

$$\delta T = L/(c_0 - v) \simeq \frac{L(1 + v/c_0)}{c_0} \tag{6.3}$$

I have taken the important step of validating the logic and the method of the experiment, and its ability to test whether the one-way relative velocity is Galilean, with *sound*, which is a familiar Galilean wave. As it is well known, *the velocity of sound is independent of the velocity of the source and depends only on the properties of the medium* (this is true for other waves and was established for light as well before the days of relativity theories). The measurement is done in the moving frame, with both the emitter and the receiver comoving on the same movable platform.

If the source and the detector are moved slowly through the medium, air, the duration that sound takes to propagate a fixed distance L changes linearly in v. For $v \simeq 1$ m/s, and $L \simeq 8.6$ m, the change in propagation duration is 73 μs, which is measurable directly in a careful experiment (there are no synchronization problems with sound; therefore the experiment is feasible in a straight path as well as in a 'stadium' configuration).

The source is a buzzer operated by a periodic short pulse (1 ms) at 4–5 kHz base frequency. The detector is a miniature microphone, fixed facing the opposite direction to receive the sound that comes to it after round trips. The folded round-trip path is defined with 5 cm diameter PVC pipes with smooth bends. The source-detector assembly can be smoothly moved over a distance of about 50 cm through a small long slot in the pipe (Fig. 6.9).

The result of the experiment is plotted in Fig. 6.9, which confirms that the round-trip duration changes linearly with the velocity of the frame (Unnikrishnan, 2018a, 2020a). As expected, the result confirms that the propagation of sound is Galilean and the result for the round-trip path is identical to the one expected in a linear path. *This validates the experimental method, freeing it from conventions, postulates and theories of relativity.*

What do we expect for light? *It is not Galilean according to the STR. So, the experimental results should be different from those for*

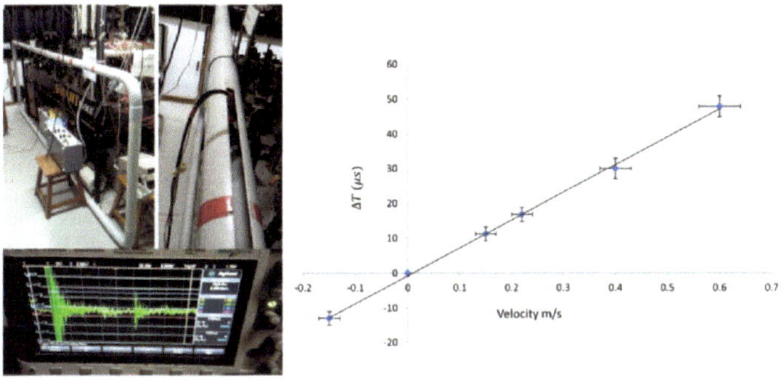

Figure 6.9 The scheme in the TIFR lab for the measurement of the relative velocity of Galilean sound along a 'stadium path'. Left: The experimental arrangement and the oscilloscope trace of the sound pulses at the source and after the round trip. Right: The measured duration of propagation of sound as a function of the velocity of the 'observer' frame, confirming its expected Galilean nature.

sound, relative to the moving frame. In the lab frame, since the path length from the source to the detector changes with velocity, as $L(v) \simeq L(1 + v/c)$, the duration also should change in a similar way, such that velocity is the invariant c. In the moving frame, the path length remains constant at L'. This is easy to see, since L' is the distance to the detector D from the source S measured along the round-trip path, in the rest frame of the source. Since there is no relative motion between S and D, L' remains fixed. So, if the relative velocity of light, $L'/\delta T'$, is an invariant constant, then the duration $\delta T'$ also should remain fixed.

However, it is a simple exercise to show that the duration according to the Lorentz Transformations is identical to that in the Galilean Transformations for this closed and piece-wise inertial path, to first order in v/c; we get $\delta T'(v) \simeq \delta T_0(1 + v/c)$. The proof does not need any elaborate calculation; for any closed path, the integral of the Lorentz term vx/c^2 is zero. Therefore, both transformations are identical to first order in v/c. We can already see that the Lorentz Transformations are inconsistent with the postulate that the physical speed of light is an invariant, when we go beyond the trivial scope of motion (inertial or non-inertial) along a linear

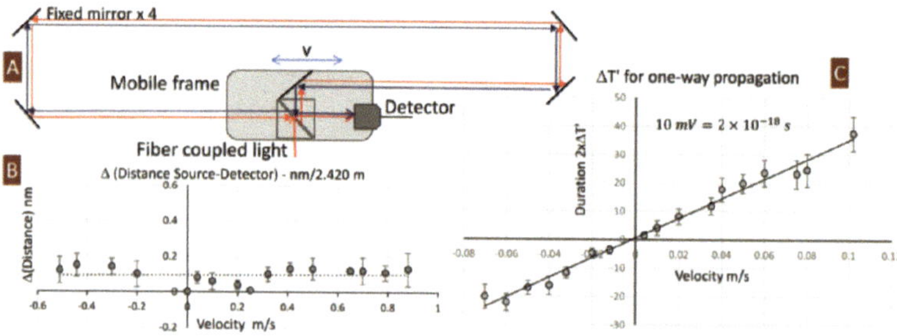

Figure 6.10 A: The optical scheme in the TIFR lab for the measurement of the dependence of the one-way relative velocity of light on the velocity of the reference frame. B: Data that establish the independence of the path length L' (relative distance) when the velocity of the platform (reference frame) is varied. C: Results showing the Galilean nature of the propagation of light. The effective relative duration is plotted. Compare with the results on the experiment with the Galilean sound.

piece of path with separated clocks. But, our focus here is what real direct experiment has to say on the relative speed of light.

Figure 6.10A shows the scheme of my experiments with light, in an identical configuration as the one we discussed for sound. The measurement, as before, is in the moving frame. The quantity measured is the difference in the durations in the two directions, measured interferometrically, as a phase difference. This way, we are measuring $\delta T_R - \delta T_L$, or the anisotropy. For a Galilean wave,

$$\delta T_R - \delta T_L = \frac{L}{c(1-v/c)} - \frac{L}{c(1+v/c)} = \frac{2Lv}{c\ c} \tag{6.4}$$

precisely because the relative velocity in the moving frame is anisotropic. For the isotropic light of the STR, with an invariant velocity, the result should be very different.

The light from a CW laser (He-Ne or a diode laser at 780 nm) is fibre coupled onto a movable platform containing beam splitters, mirrors and a split photodiode detector. Two identical round-trip paths are created with the beam splitter and mirrors (Fig. 6.10A), with long straight sections such that the moving frame is inertial during the measurement (supplementary experiments clearly show that this restriction is not necessary, but there is no need to discuss those aspects here). The total path length is about

3 m. The alignment is done by shifting the platform to various positions and making sure that the fringe pattern remains stable at all positions (with the speed of motion zero). For low velocity, 0.1 m/s, the expected additional one-way delay, if (and only if) light is Galilean, is 3.3×10^{-18}s. This corresponds to about 1/1000 of the wavelength and it is not difficult to measure, once the alignment remains good and the noise due to the vibrations during the movement of the platform is controlled. (The sensitivity can be increased with a longer path length. The use of optical fibres can enormously increase the signal, but if the propagation is not in free space, the corrections for the Fresnel drag, etc., bring in some theoretical overhead.) It was not possible to do the measurements at velocities larger than about 10 cm/s, because of the blurring of the fringes due to the vibrations in the platform. (This situation was improved later and a new interferometry experiment with femtosecond laser pulses is in progress in our laboratory, employing pulse detection techniques feasible only with fast high-power pulses. The results are consistent with the previous measurements. The direct measurement of the propagation delay, without using interferometry, is difficult for a short path length of the order of few metres, even with femtosecond pulses.) The fact that the path length L' remains constant (independent of the velocity of motion) to first order in v/c is verified by supplementary measurements of the two-way propagation of light, by retro-reflecting light and monitoring the stability of the interference fringes (Fig. 6.10B).

The data from the experiment, plotted in Fig. 6.10C, follow eq. 6.4 for Galilean waves, and the results go against the expectation of the STR (Unnikrishnan, 2020a). The duration for the one-way propagation of light, measured in the moving frame, varies linearly with the velocity of the frame, $\delta T \simeq Lv/c^2$, as for the Galilean sound! By comparing the behaviour of the two kinds of waves, expected to be widely different, we have eliminated uncertain theoretical and interpretational factors. The result unambiguously establishes that the relative velocity of light is Galilean, $c \pm v$, and not an invariant. *The fundamental postulate of the Special Theory of Relativity stands falsified by the results and with that, the theory itself falls apart.*

The experimental result reiterates Einstein's simultaneity fallacy we encountered in the previous section. Though the experimental result predates my noticing the fallacy, it is very satisfactory that the two facts go together. The observed perception of the violation of absolute simultaneity and the Galilean propagation of light necessarily go together.

This result clearly demands another theoretical paradigm with a privileged (absolute) frame that predicts and agrees with all known empirical evidence, but very different in the foundations and content from the STR. We have already seen the salient features of that paradigm and the theory of Cosmic Relativity. In fact, *the Galilean nature of the propagation of light was the primary prediction of Cosmic Relativity*. We will now see the action of cosmic gravity on clocks in a spectacular application of Cosmic Relativity – the Global Navigational Satellite Systems like the Global Positioning System (GPS).

6.4 Summary

This interlude presented two key proofs that invalidate Einstein's Special Theory of Relativity. One was a cardinal error made by Einstein in analysing the simultaneity of spatially separated events and the local synchronization of clocks. While he assumed the non-Galilean constancy of the relative velocity of light in all inertial frames, his conclusion on simultaneity of distant events was that of Galilean physics, proving the internally inconsistency of the STR. What is ironic about this error is not only that it is elementary enough to be analysed and understood easily, but also was pointed out to Einstein by H. Bergson, both in person and in a monograph, *Duration and Simultaneity* (Bergson, 1922). Einstein never tried to counter directly Bergson's transparent demonstration. Now, I have provided the solid proof that Einstein was indeed in error, by comparing the simultaneity of a pair of events employing both Galilean messenger waves (like sound) and Einstein's 'non-Galilean light', with relevant logical and algebraic details. Each can see for oneself, step by step, and be convinced of the internal conflict in Einstein's STR.

The second vital proof consisted of the *experimental results* on the nature of the genuine one-way propagation of light, and a comparison with the propagation of the Galilean waves of sound in identical experimental configurations. The results show clearly the Galilean nature of the propagation of light, with its variable velocity relative to an inertially moving frame being $c \pm v$. Irrespective of any theoretical prejudice, if the relative velocity of light is measured to be $c \pm v$, it is Galilean and Einstein's Special Theory of Relativity is *falsified in toto*.

One might then ask, in disbelief, why so many experiments in the past were in agreement with the STR. This is true as a fact, and it is an equally true fact that those very experiments are also in perfect agreement with Cosmic Relativity, and even with the ether-based relativity of Lorentz and Poincaré as well! There is a class of experiments that are predicted to have the same results in the STR, which denies any absolute frame, as well as in theories of relativity with an absolute frame, like the LPTR or Cosmic Relativity. Experiments that deal with the two-way propagation of light and second-order (v^2/c^2) effect are in this class. So are the experiments on the propagation of light that use spatially separated clocks in the measurements. The falsehood of the theory becomes visible only when crucial experiments of the right kind are performed. Some experiments in this crucial class deal with the one-way propagation of light without spatially separated clocks, electrodynamical effects with moving conductors and magnets, clock comparisons in certain round-trip configurations, etc. What is important to note is the basic dictum of physics: *even if scores of experiments are in agreement, a single crucial experiment in disagreement falsifies the theory*. Finally, anybody can look through the mighty telescopes and be convinced of the vast amount of matter in the Universe. The matter we observe has gravity. With so much matter and energy in the physical environment, a theory that entirely depends on the emptiness of space for its basic axioms to be even approximately true, cannot be correct.

Then, what impossible miracle can one hope for... and wait for, to save the falsified theory? There is nothing at all to save an empty space paradigm in this matter-filled gravitational Universe. The unbearable heaviness of carrying the burden of the inconsistent theory will become even more evident when we study the motional modifications of the clocks of satellite-based navigational systems like the GPS, and their signal transfer protocol and algorithms that depend on the nature of the propagation of light.

Chapter 7

The Moving Clocks of Navigational Satellite Systems

The Global Navigational Satellite Systems (GNSS) based on a network of satellites around the Earth, like the GPS, rely on precision atomic clocks and radio signals to locate a 'receiver' in space, typically at or near the Earth's surface. GLONASS (Russia), Galileo (Europe), and BeiDou (China) are other global systems. There are also regional navigational systems, such as the IRNSS/NavIC (India) and the QZSS (Japan). These high-precision navigational systems with good 'civilian access' are technologically and socially transformational. The basic idea that relies on the light hypothesis is simple. If the relative velocity of light does not depend on the velocity of either the source or the observer, the observer is equivalent to a stationary location, with the rest of the world moving relatively. Then, the duration to propagate from the source to the observer, when multiplied by the constant c, is equal to the distance between them, irrespective of any motion of the observer. Thus, GPS-like systems rely on the hypothetical equivalence of space and time (Ashby, 2002, 2003).

However, we saw in the previous chapter that the light hypothesis clashes directly with the conclusions on the relativity

Gravity's Time
C. S. Unnikrishnan

ISBN 978-981-4877-72-5 (Hardcover), 978-1-003-25656-4 (eBook)
www.jennystanford.com

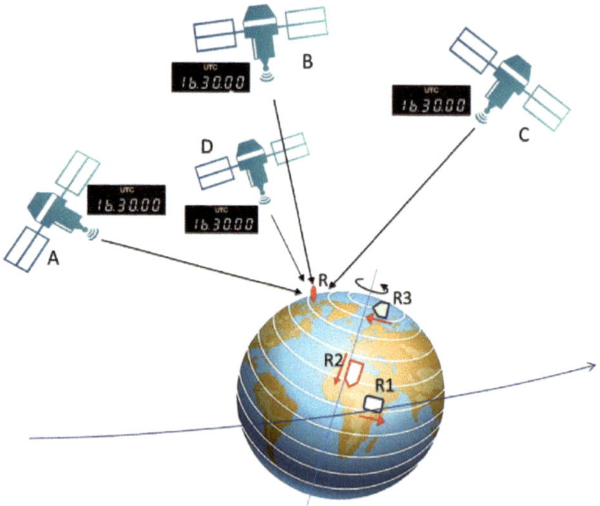

Figure 7.1 The typical GNSS configuration. An accurate time signal t_s is received from each satellite at the receiver R at time t_R, which then has several records of time differences. This is converted into distances, and then the position of the receiver is calculated. The Earth's rotation and its orbit around the Sun are indicated. Each of the mobile platforms R1, R2, R3 in inertial motion can get their dynamical position.

of simultaneity. How can the GNSS operate correctly with an inconsistent theory backing it? Evidently, they cannot. It turns out that these systems are factually calculating the location using the Galilean propagation of light and the motional time dilation in a theory of relativity with an absolute frame, instead of the tenets of the STR! I will go through the details of timing with the atomic clocks in the GNSS to demonstrate that this is indeed the case.

Suppose there are many moving sources of accurate time, distributed in space (Fig. 7.1). These sources A, B, C. . . , etc., send the digital records of their time and their locations, which are accessed by the receiver R at time t_R to make record of time differences $\Delta t_A = t_A - t_R$, $\Delta t_B = t_B - t_R \ldots$, etc. *The only measured quantity is the difference in time between the emission and reception of the signal.* This is done by moving a pattern of 1 and 0 in the receiver until it matches with the unique pattern sent from the satellite. The details are not important for understanding how the motion and gravity affect the time in the GPS navigation. Now the theory kicks in. Assuming the invariance of the relative speed of light c in the STR, an algorithm finds the several 'ranges' $d_A = c\Delta T_A$, etc., from the product of the constant speed of light and time differences. These are the distances at t_R from R to the sources A, B, etc., as R 'sees' them, at their positions when they transmitted their time codes. If one has three such ranges, each of them defines a sphere in space such that the receiver could be anywhere on the surface of that sphere, relative to the particular source. The intersection of two spheres is a circle, and the intersection of that circle with a third sphere is just two points. The unique point defining the position of the receiver can be picked from this because the satellites are at a large distance and only a portion of the curved surface of the 'spheres' are to be considered. They are more like intersecting patches of curved surfaces; then the intersection of two is a curved line, instead of a whole circle, and its intersection with another surface is a unique point in space. Allowing for small timing error, one can locate a small region in space. This is the principle of 'trilateralization' by which a GNSS can locate a receiver. But, since the clock in the receiver (say, in a mobile phone) is not a high-precision atomic clock, the system needs one more equation for the accurate calculations. Therefore, four, instead of three, ranges to the satellites equipped with the atomic clocks are required for precise localization.

There are high-precision reference clocks, which are state-of-the-art atomic clocks, at the Earth's surface, in specified locations, at relative rest in the laboratories and stations. The clocks in the satellites are also atomic clocks, but with modest accuracy and stability, sufficient to serve the purpose of localization with millimetre accuracy, if needed. A clock that has a basic stability of 10^{-13} keeps time to an accuracy of 10 ns in a day. The distance travelled by light in 10 ns is 3 m. The clocks in the satellites need to be corrected periodically using the more stable reference clocks, if the ranging accuracy of a few centimetres has to be maintained.

The duration of propagation indicated by the difference between the satellite and receiver clock readings needs to be corrected for several physical effects. The fact that the radio waves have to pass through the atmosphere of the Earth, which contains significant density of charged particles (ionosphere), results in a considerable propagation delay and navigational error (several metres) that need to be corrected for, by using active models of the slowly varying electron density in ionosphere, using extensive measurements that are available.

The gravitational time dilation of the clocks in the Earth's gravitational field is a large relativistic effect to be corrected in the GNSS. The satellites are at a significant distance from the surface, about 20000 km for the GPS and about 36000 km for geosynchronous systems such as the Indian Regional NSS. Hence, they are at a lower gravitational potential, compared to the clocks at the surface of the Earth (at the mean sea level). While the potential at the surface is $\phi_0 \simeq -GM/R_e$, the potential at the satellite at the distance r_s is $\phi_s = -GM/(R_e + r_s)$. The Earth's radius is approximately 6400 km, which is much smaller than the distance to the satellite. Therefore, we cannot use the approximate formula for time dilation, $\delta T/T \simeq -gd/c^2$. The gravitational potentials at the locations of the clocks have to be used for calculating the gravitational time dilation. The relative gravitational time dilation is

$$\Delta T_g/T_g = (\phi_0 - \phi_s)/c^2 = \frac{-GM}{c^2}\left(\frac{1}{R_e} - \frac{1}{R_e + r_s}\right) \tag{7.1}$$

This time dilation factor, if not corrected, will cause considerable errors on the determination of locations, making navigation impossible. For the GPS, this is about 17 cm error in a second. Therefore, the navigation will be off the road in about 10 s, unless a correction is applied to the clocks in the satellites, to bring them in tune with the reference clocks on the Earth. They are corrected well in the pre-tuning of the rate of the clocks as well as in the GNSS software. This actual corrections take account of the irregularities in the potential due to the Earth's irregular topography and sub-surface structure, etc., using the measured data. The small variations in the potential due to the different positioning of the receivers near the Earth's surface is not significant enough for navigational errors. Even for an aircraft at 10 km height, the gravitational time dilation relative to a clock at the Earth's surface is only $\delta T/T \simeq gh/c^2 \simeq 10^{-12}$, which translates to only 3 mm error in 10 s.

So, we conclude that the correction to compensate for the gravitational time dilation, due to the variation of the Earth's potential with the height, is done correctly and adequately in the GNSS like the GPS and the GLONASS. The gravitational potential due to the farther and more massive sources such as the Sun, Milky Way, etc., though larger, is more or less the same at the position of the satellite clocks as well as at the location of the clocks on the Earth's surface. Therefore, the *relative time dilation* due to these massive structures is negligible.

But, the situation regarding the corrections for the motional time dilation in the GNSS is bizarre, to say the least, and blatantly non-STR as we will see now.

7.1 Comparing Clocks, Revisited

The only practical way of comparing clocks globally is by sending electromagnetic signals, either through space or through optical fibres. Then, our understanding of the propagation of light becomes crucial. At present, all clock comparisons in important applications such as the GNSS depend on the postulate of the invariance of the relative speed of light in the Special Theory of Relativity, or the *light hypothesis*. However, since there is no independent verification of the postulate by any experiment, one has to depend on consistency checks. If all is well when the light hypothesis is used, then one can have confidence in the theory and its application. Otherwise, obviously, the theory and the protocol of clock comparison need to be changed.

As I mentioned on different occasions, the protocol for clock comparison and range finding with Einstein's light hypothesis is simple. Before the launch of the satellite, the clocks are synchronized with the reference clocks on the Earth. The changes that happen in the launch and the motion of the satellite need to be corrected once, and this can be done by the two-way communication between the satellite and the earth-stations. Then, we can assume that the satellite clocks are synchronized to the reference clock on the Earth, at some reference 'zero' of time. Subsequently, the relativistic time dilation effects that alter the synchronization need to be corrected for regularly. If the corrections are applied promptly, then the clocks remain synchronized. If a light pulse takes the duration ΔT to reach from the satellite to a receiver at a location on the Earth's surface, the propagation distance (range) *in the rest frame of the receiver* is given simply as $D = c\Delta T$, according to the STR. This is always independent of the motion of the receiver as per Einstein's theory.

On the other hand, if the relative velocity of light is Galilean and varies with the velocity of the observer (receiver), the range equation should be different, $D(v) = c(v)\Delta T$. At this stage, I emphasize that only the quantity 'duration', the difference of the reading of the clocks, is measured and used in the GNSS systems. The calculation of the range relies entirely only on ΔT. Therefore, if any velocity dependence is *empirically* seen in actual GNSS ranges, it has to be implemented through a velocity dependence of ΔT, by rewriting the equation as $D(v) = c\Delta T(v)$. That is, if $c(1+f(v))$ is the needed correction for the relative velocity of light, the factor $1+f(v)$ needs to be associated with the propagation delay as $\Delta T(1+f(v))$, instead of with c, because the velocity is already assumed to be an invariant constant, as per the STR; the GPS system deals only with the vocabulary of time, and not with the velocity. This observation is relevant and important, as we shall soon see.

7.1.1 Motional Time Dilation

The typical GNSS systems use satellites that go around twice a day (semi-diurnal), at a distance from the Earth's centre of about 26000 km. The orbital speed of such satellites v_s is approximately 3.9 km/s. This is the speed of the Keplerian orbit of the satellites, $v_s = \sqrt{GM/R_s}$. However, what is relevant for the time dilation in the STR is the relative speed, with reference to the receiver (observer) on the Earth's surface, which may have a motion with the ground velocity v_r. Since the satellite is moving relative to the receiver, at the relative speed $v_{s(r)} = \vec{v}_s - \vec{v}_r$, the motional time dilation of its clocks, according to the STR, is

$$\begin{aligned}\Delta T_m/T_m &= -\frac{v_{s(r)}^2}{2c^2} = -\frac{(\vec{v}_s - \vec{v}_r)^2}{2c^2} \\ &= -\frac{v_s^2}{2c^2} - \frac{v_r^2}{2c^2} + \frac{\vec{v}_s \cdot \vec{v}_r}{c^2}\end{aligned} \tag{7.2}$$

If the receiver is at rest near the equator, it shares the Earth's rotation at the speed of $v_r = 0.464$ km/s to the east. Since the satellites are in orbits of different inclinations relative to the Earth's rotation axis, the relative velocity $v_{s(r)}$ is not a fixed quantity. However, with $v_s \pm v_r$, one can see that the motional time dilation $-v_{s(r)}^2/2c^2$ could range between 6.6×10^{-11} to 1×10^{-10}. This amounts to a range error of $c\Delta T_m$ about 1 m in 40 s.

In contrast, if Cosmic Relativity is the factual relativity, and if the cosmic frame is the absolute master frame, then the expression for the motional time dilation will be very different. We have to calculate the time dilation of each clock using its speed of motion relative to the cosmic frame and then take their difference to get the relative time dilation. However, the motional time dilation that is common to the clocks does not appear in the final expression for the difference, when we assume that $v \ll c$. Since the satellites are in their orbits around the Earth, they are moving with the orbital motion of the Earth, and so are the receiver clocks. So, the Earth's orbital velocity and larger contributions (motion of the solar system, etc.) drop out of the expression for the time dilation. If v_s is the Keplerian velocity of the satellite relative to the Earth's centre, the time dilation of its clocks is $\delta T_s/T_s = -v_s^2/2c^2$. Similarly, if v_r is the velocity of the receiver relative to the same reference point, its time dilation is $\delta T_r/T_r = -v_r^2/2c^2$. Then, the relative motional time dilation is

$$\Delta T_m/T_m = -\frac{v_s^2}{2c^2} - \left(-\frac{v_r^2}{2c^2}\right) = -\frac{v_s^2}{2c^2} + \frac{v_r^2}{2c^2} \qquad (7.3)$$

This is very different from the STR formula 7.2; the sign of the second term is opposite, and the crucial bilinear term $\vec{v}_s \cdot \vec{v}_r$ that is characteristic of a relative velocity is not present. When the velocity of the receiver is the typical speed of a terrestrial vehicle or even an aircraft, $v_r \ll v_s$. Then the second term is much smaller than the first, and the motional time dilation in the absolute relativity theory is expected to be a relatively fixed $v_s^2/2c^2$, which amounts to about 8.5×10^{-11} s for every second in the clocks of the GPS. The difference from the STR is easily distinguished.

7.1.2 Range Finding

It is in the algorithm to calculate the position of a receiver, which is the relativistic core of the GNSS, that we find clear evidence for the reality of an absolute reference frame of motion. *The algorithm that works, and is in use, is what is prescribed by relativity with a master frame, and not the one conforming to the STR.* In particular, the use of a correction to the time record to compensate for the Galilean propagation of light, with a non-constant relative velocity that depends on the velocity of the receiver (target), is revealing and striking.

Let us follow the GNSS navigational protocol. The nth satellite sends a timing signal to the GNSS receiver at some local time t_n, as per the satellite clock. This is received at t_r at the receiver. The difference in time between sending and receiving the signal is $t_r - t_n$. Since the STR assumes that the speed of light does not depend on the speed of either the satellite or the receiver, the distance from the receiver to the satellite, or the range, is

$$D_s = c(t_r - t_n) = c\Delta t_{rn} \tag{7.4}$$

This is the defining formula of the GNSS design, by which the localization of a target is done, enabling navigation. The duration of propagation from the satellites at 26000 km is about 0.087 s, or 'half a blink'. There is absolutely no dependence on the velocity of the satellite or the receiver in this formula, as expected in the STR. However, this is not what is used in the GNSS! Additional corrections were incorporated after seeing that the STR range formula did not serve the purpose (Ashby, 2002).

In contrast to the situation in the STR, if the theory of relativity is based on a privileged master frame, like the matter-filled Universe, the range equation will be very different. In the rest frame of the receiver, the relative velocity of light can vary from $c + v_r$ to $c - v_r$, resulting in a dependence of the range on the velocity of the receiver.

The range equation assuming the Galilean propagation of light becomes

$$D_s = c'(v_r)(t_r - t_n) = |(\vec{c} - \vec{v}_r)|\,\Delta t_{rn} \tag{7.5}$$

The magnitude of the relative speed of light is then

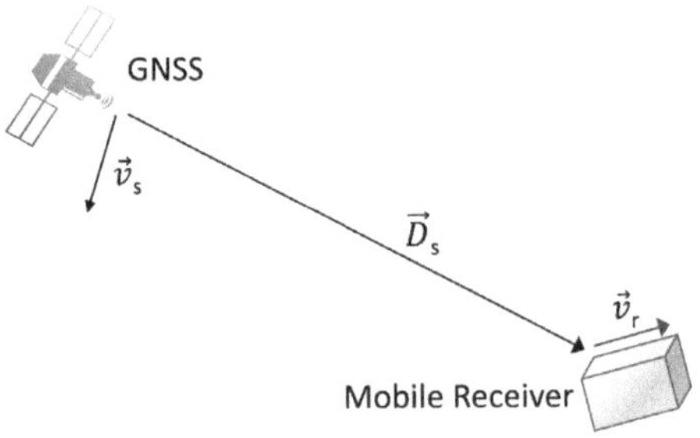

Figure 7.2 The typical navigational configuration. In the frame of the mobile receiver, the STR prescribes that the relative time dilation is $\delta T/T = -v_{sr}^2/2c^2$, where $v_{sr} = |\vec{v}_s - \vec{v}_r|$. In sharp contrast, Cosmic Relativity predicts $\delta T/T = -v_s^2/2c^2 + v_r^2/2c^2$. In the STR, the relative distance (range) is just $D_s = cT_{sr}$, where T_{sr} is the time of propagation as marked by the clocks in the satellite and the receiver. In Cosmic Relativity, the variable relative velocity of light translates to an additional correction for the range, $D_s' = D_s + \vec{D}_s \cdot \vec{v}_r/c$. This is equivalent to the corrected duration, $T_{sr}' = D_s'/c = T_{sr} + \vec{D}_s \cdot \vec{v}_r/c^2$.

$$|(\vec{c} - \vec{v}_r)| = \sqrt{(c^2 + v_r^2 - 2\vec{c} \cdot \vec{v}_r)} \simeq c(1 - \vec{c} \cdot \vec{v}_r/c^2) \tag{7.6}$$

where we have assumed that $v \ll c$. Since $\vec{c}(t_r - t_n) \simeq \vec{D}_s$, the modified range equation is

$$\begin{aligned} D_s &= c'(v_r)(t_r - t_n) \\ &= \Delta t_{rn} c(1 - \vec{c} \cdot \vec{v}_r/c^2) \simeq c\Delta t_{rn} - \vec{D}_s \cdot \vec{v}_r/c \end{aligned} \tag{7.7}$$

The first term alone is what one would expect if the STR were right. In contrast, there is a characteristic extra correction proportional to the velocity of the receiver, necessary in the theory of Cosmic Relativity (Fig. 7.2). This is simple to understand; if the receiver happens to be moving away in the cosmic frame as the radio-waves arrive with the clock signal, the relative velocity of the waves is less than c, and it takes more time to reach the receiver. If the receiver is moving towards the waves, the relative velocity is more than c, resulting in less duration of propagation. Even for a receiver at rest on the Earth's surface, near the equator, the relative velocity of radio-waves would be about 460 m/s different from c. Then, the (maximum) time of propagation between the satellite at the horizon and a receiver would be about $D_s v/c^2 \simeq 135$ ns different from the STR value. Thus, the relativistic correction to be applied in the algorithm to get the correct range to the GNSS satellites is very different in the STR and Cosmic Relativity. This is not specific to the velocity due to the rotation of the Earth – even a fast inertial projectile or a vehicle moving near the surface of the Earth will have the same Galilean correction.

The formulae that are factually used for the motional corrections of time in the GPS and in the other GNSS are discussed in many documents (Ashby, 2003). However, they do not correspond to the STR. They are in fact exactly the ones derived in Cosmic Relativity! The range $D_s = \Delta t_{rn} c(1 - \vec{c} \cdot \vec{v}_r / c^2)$ is the product of two terms – the relative speed of light and the duration of propagation. However, the GNSS algorithms work with the timing signals, and do not deal with the relative velocity of the signal, having already assumed it to be the constant c. Therefore, the Cosmic Relativity correction to the relative velocity of light, $|(\vec{c} - \vec{v}_r)| \simeq c(1 - \vec{c} \cdot \vec{v}_r / c^2)$ is directly applied to the clocks, instead of applying it as a correction of the speed of light. That is, we can write the range as $D_s = c\Delta t'_{rn}$, with a modified time

$$\begin{aligned} \Delta t'_{rn} &= \frac{\Delta t_{rn}}{(1 - \vec{c} \cdot \vec{v}_r / c^2)} \simeq \Delta t_{rn}(1 + \vec{c} \cdot \vec{v}_r / c^2) \\ &= \Delta t_{rn} + \vec{D}_s \cdot \vec{v}_r / c^2 \end{aligned} \tag{7.8}$$

(The correction on the duration of propagation has the sign opposite of the correction on the velocity of propagation. The lesser is the relative velocity, the more is the duration of propagation.) In other words, since the actual relative speed of light is $\vec{c} \pm \vec{v}_r$ the duration of propagation over the distance D_s is

$$\begin{aligned} \Delta t' &= \frac{D_s}{|\vec{c} - \vec{v}_r|} = \frac{D_s}{\sqrt{c^2 + v_r^2 - 2\vec{c} \cdot \vec{v}_r}} \simeq \frac{D_s}{c}(1 + \vec{c} \cdot \vec{v}_r / c^2) \\ &= \frac{D_s}{c} + \frac{\vec{D}_s \cdot \vec{v}_r}{c^2} \end{aligned} \tag{7.9}$$

When the range is expressed as $D_s = c\Delta t'_{rn}$ the correction that is factually due to the Galilean propagation of light is ascribed to the time of the clocks.

Summarizing, the STR formula for the motional time dilation is $\Delta T_m / T_m = -v_{s(r)}^2 / 2c^2$, where $v_{s(r)}$ is the relative velocity of the satellite, relative to the GPS receiver, and the range formula is $D_s = c\Delta t_{rn}$. One then expects that the GNSS implement these motional relativistic factors faithfully, confirming the trust in the light hypothesis and in the STR. Not at all! This is not the correction for the motional time dilation that is factually applied to the GNSS clocks. The range equation is not the STR equation either. The GNSS algorithms follow faithfully the absolute frame relativity, as prescribed by Cosmic Relativity, *because that is the only prescription that works* for localization and navigation. Since this is an extraordinary fact, let us examine it closely.

7.2 The Factual Motional Relativity in the GNSS

We have already seen that the relative time dilation of a clock 's' relative to another clock 'r', in the absolute master frame of the cosmic gravitational paradigm, is

$$\frac{\Delta t'}{t'} = \frac{dt'_s}{t'_r} - \frac{dt'_s}{t'_r} \simeq \frac{-v_s^2}{2c^2} - \frac{-v_r^2}{2c^2} \qquad (7.10)$$

In sharp contrast the motional correction in the STR is

$$\frac{\Delta t'}{t'} \simeq -v_{sr}^2/2c^2 \qquad (7.11)$$

where v_{sr} is the relative speed of the satellite (A), relative to the rest frame of the receiver (B) clock. Further, there is a 'first order' (proportional to v_r) correction in Cosmic Relativity that is due to the variable relative velocity of light, which depends on $\vec{v}_r$, the distance $\vec{D}_{sr}$ between the clocks, and the angle between the two vectors (Fig. 7.2). It is

$$D'_{sr} = c\Delta t_{sr} + \vec{D}_{sr} \cdot \vec{v}_r/c \qquad (7.12)$$

to be compared to the STR formula $D_{sr} = c\Delta t_{sr}$. The range correction in the second term, when applied to the clocks, is

$$\delta t_{sr} = \vec{D}_{sr} \cdot \vec{v}_r/c^2 \qquad (7.13)$$

The sign corresponds to choosing the positive direction of $\vec{D}_{sr}$ along the propagation of the radio-waves, from the GNSS satellite to the receiver.

The factual corrections implemented in the GNS systems such as the GPS, GLONASS, and GALILEO indeed conform to the 'absolute algorithm', contained in the Cosmic Relativity formulae in eqs. 7.10 and 7.12. They are not based on the STR formulae!

The need for this change from the simple STR formulae was realized after the GNSS programs started; naturally this was not anticipated within the STR. However, the change was implemented because of the empirical need. We should understand how the GPS scientists deal with these corrections and how they try to interpret them, in an effort to reconcile the conflict with the STR (Ashby, 2002, 2003). The non-STR formula for relative time dilation was given a justification, 'after the act', by decreeing that one should not use the rest frame of the receiver. Then, of course one is not using the STR and its relative velocity to calculate motional corrections. It was arbitrarily stated that the slow rotation of the Earth makes the receiver rest frame 'non-inertial', which necessitates the choice of a more global 'Earth-centric non-rotating frame' for all the calculations. It is simple logic to realize that the same (false) argument cannot be used in the case of a fast inertial mobile moving near the Earth, not sharing its rotation! Therefore, it is straightforward to conclude that the GNSS operation falsifies the STR. This becomes even clearer when we examine some more details.

(It is perhaps relevant to point out that there is a prevalent misunderstanding that the STR cannot be used in non-inertial frames. The same excuse was given by Hafele and Keating when they encountered unexpected results in their clock comparison experiment. If that were true, one would have to conclude that the STR was not valid for most situations in the electrodynamics of charged particles, including atomic physics. Generations of students learn atomic spectra and the relativistic factor of Thomas precession by applying the Lorentz Transformations of the STR in the very non-inertial rest frame of the electron. The starting point of Einstein's discovery of the General Theory of Relativity was the assumption of the applicability of the STR in accelerated frames. It is high time that one refrains from such unphysical reasoning, when faced with unexpected results.)

The range correction equation 7.12 is general, and accurate enough for most situations. As we discussed earlier, the only measured quantity in the GNSS is the difference of time marked by the clocks at different positions and velocities. All corrections have to be expressed as a correction to the clocks. We saw that the extra first-order correction to be applied to the clocks in the GNSS is

$$\delta t_{rn} = \vec{D}_s \cdot \vec{v}_r / c^2 \quad (7.14)$$

By transferring the excess delay due to the Galilean propagation as a forced correction to the clocks, one can continue to pretend that the relative speed of light is the invariant c, when it is not.

It is important to note that even for a GNSS receiver at rest on the Earth's surface, v_r is not zero; there is the local rotation speed of the Earth $\vec{v}_r = \vec{\Omega} \times \vec{r}$ to be reckoned in its absolute velocity in the cosmic frame. This is about 460 m/s at the equator and progressively smaller at the higher latitudes. We came across this during the discussion of the Hafele–Keating experiment. For this special case of the receiver at rest or very small velocity,

$$\vec{D}_s \cdot \vec{v}_r = \vec{D}_s \cdot \vec{\Omega} \times \vec{r} = \vec{r} \times \vec{D}_s \cdot \vec{\Omega} \quad (7.15)$$

This is maximum at the equator, and zero at the poles. Figure 7.3 explains this correction graphically.

However, the GNSS scientists do not understand it this way, because typically the velocity of the receiver is dominated by the rotational velocity of the Earth, which a slowly moving GPS receiver shares. We saw that the expression $\vec{r} \times \vec{D}_s \cdot \Omega$ came from $\vec{D}_s \cdot \vec{v}_r$, with $\vec{v}_r = \vec{\Omega} \times \vec{r}$. The GPS physicists write $\vec{r} \times \vec{D}_s \cdot \vec{\Omega}$ as $\frac{1}{2}\vec{r} \times \vec{D}_s \cdot 2\vec{\Omega}$, and then identify $\frac{1}{2}\vec{r} \times \vec{D}_s$ as the area A of the triangle connecting the satellite, the receiver, and the centre of the Earth. Writing the correction as $2\vec{\Omega} \cdot \vec{A}/c^2$ suits the misinterpretation that the correction is the Sagnac effect, which is specific only to rotating frames. However, as I stressed, this is a special case of the general range correction $\vec{D}_s \cdot \vec{v}_r / c^2$, and it is necessary for the navigation of mobile platforms that do not share the rotation of the Earth as well. This fact shows that the correction is factually due to the Galilean propagation and not due to any non-inertial feature of the motion of the receiver. Ironically, the Sagnac effect was the anisotropic phenomenon of the propagation of light in a rotating interferometer, highlighted by G.

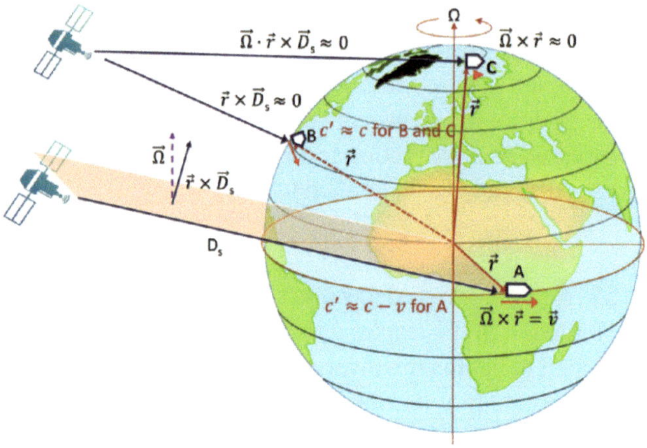

Figure 7.3 The deconstructed physics and geometry of the GNSS range calculation. The propagation of light with the relative velocity $\vec{c} - \vec{v}_r$ drastically modifies the range equation. For targets at rest on the Earth, the velocity v_r due to the rotation of the Earth depends on the latitude. Only the component of the velocity along the propagation of the GPS signal is relevant for the relative velocity of light and the range correction $\vec{D}_S \cdot \vec{v}_r = \vec{D}_S \cdot \vec{\Omega} \times \vec{r} = \vec{r} \times \vec{D}_S \cdot \Omega$. The target A needs the maximum correction. The target B needs a negligible correction because its velocity is nearly perpendicular to the direction of propagation. The target C needs a negligible correction because the velocity is small near the poles.

Sagnac in 1913, to prove that the ether existed and that the relative velocity of light was Galilean! In that sense, the GPS interpretation of the clock correction invoking the Sagnac effect is spot-on, though inadvertent, delivering scientific justice to Sagnac (Sagnac, 1913a; Unnikrishnan, 2020b).

In fact, the widespread confusion between the time delays in the propagation of light and the factual changes in the reading of physical clocks is reflected in the official descriptions of this extra correction in the GPS. To quote the GPS physicist Neil Ashby (Ashby, 2003),

> It is remarkable how many different ways this can be viewed. For example, from the inertial frame it appears that the reference clock from which the synchronization process starts is moving,

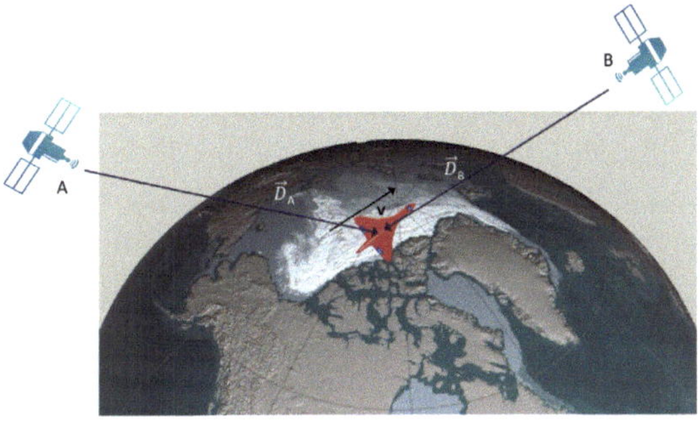

Figure 7.4 A fast inertial flight in its polar route, guided by the GNSS. For the time code from the satellite A, the correction is $\vec{D}_A \cdot \vec{v}/c^2$ and for the satellite B, the correction is of the opposite sign, $\vec{D}_B \cdot \vec{v}/c^2$. These corrections are evidently not related to the rotation of the Earth.

> requiring light to traverse a different path than it appears to traverse in the rotating frame. The Sagnac effect can be regarded as arising from the relativity of simultaneity in a Lorentz Transformation to a sequence of local inertial frames co-moving with points on the rotating earth. It can also be regarded as the difference between proper times of a slowly moving portable clock and a master reference clock fixed on earth's surface.

There is only one way to view the corrections, factually. The only reality is the arrival of the timing signal from the satellite to the rest frame of the receiver, brought by radio waves. There is only one observer – the one with the receiver. There is no one observing from another inertial frame. There are no clocks that are slowly transported from the location of the receiver to the satellite. The Sagnac effect is a Galilean result that needs an absolute frame, as shown in Fig. 6.7, and as also derived in Michelson's paper cited in Fig. 6.6. It cannot be obtained from a sequence of Lorentz Transformations around a rotating frame; that would be just another Lorentz Transformation plus a rotation, resulting in effects such as the Thomas precession, and not the Sagnac effect!

I emphasize again that the range correction could be used in this 'Sagnac form' (as the vector product of an angular velocity and an area) only in the special case of slow mobiles on the surface of the Earth, sharing the Earth's rotation. For a fast inertial vehicle, such as a supersonic flight on a polar route, the rotation of the Earth is not relevant and there is no 'Sagnac effect'. But, the Cosmic Relativity correction with the formula 7.12 still applies. This important fact is depicted in Fig. 7.4. For a flight at the supersonic 'Concorde' speed, this correction could be about 100 ns, in each location update in the GPS, amounting to more than 30 m navigational error. That is because the relative velocity of light differs from the value c by up to 600 m/s inside a supersonic vehicle.

We see that this clock correction, along with the second-order correction for the rate of the observer's clock in the rest frame, constitute undeniable evidence that falsifies the light hypothesis and Einstein's theory. I particularly emphasize the point that adequate corrections are indeed applied to the GNSS clock algorithms now, but the correct physical reasons that underlie those corrections are evidently the Galilean propagation of light and the motional time dilation that depends on the velocity in the cosmic frame. Thus, the Global Navigational Satellite Systems are one's everyday 'pocket proof' for the Galilean nature of the propagation of light and the existence of a master frame.

The factual corrections in all the GNSS installations are justified only in terms of the cosmic gravitational time dilation of the clocks in motion, as well as with the Galilean propagation of light. The current interpretations based on the STR are obviously convoluted, and they cannot be maintained. The situation is similar to the beliefs in the geocentric planetary models with epicycles, which admittedly could explain many observed features, when backed by elaborate, but ad hoc reasoning. However, as explained in detail here, the clear evidence for the gravitational paradigm and Cosmic Relativity is overwhelming in the GNSS. Hence, the correct basis on Cosmic Relativity will necessarily prevail, eventually.

7.3 Bye, Local Time

The idea that motion could modify length and time came into physics with Lorentz's attempt to reconcile the Principle of Relativity with the hypothetical privileged frame of the ether. Lorentz had introduced the concept of local time in ether-based relativity through the expression

$$t' = t - \vec{v} \cdot \vec{x}/c^2 \tag{7.16}$$

to account for the first-order (v/c) invariance of the electromagnetic phenomena. With the correction $\delta t = -\vec{v} \cdot \vec{x}/c^2$, time in a moving frame became location dependent, as well as velocity dependent. However, Lorentz imagined it as a mere mathematical device, and not as a real transformation of the physical time. Since we cannot know or verify the time of a clock that is at some distance, one could make such a hypothesis freely to preserve the mathematical invariance of the theory. But, the physical time was still considered 'absolute' in the Galilean sense, $t' = t$. Even when the full Lorentz Transformations became known through Poincaré's work, in which the second-order dilation of duration $dt' = dt\sqrt{1 - v^2/c^2}$ was understood as a real physical effect, the vx/c^2 term was thought to be just mathematical. The velocity here was taken to be the velocity relative to the hypothetical ether.

Einstein interpreted the Lorentz Transformations very differently, treating the full Lorentz Transformations as literally physical. In contrast, the velocity in these expressions were interpreted as the relative velocity between the different reference frames, which is a kinematical notion that varies with the arbitrary choice of the reference frame, without any physical reality. Yet, a seemingly consistent and successful theory emerged.

But, *the physicality of local time collapses as soon as we consider the possibility of clock comparisons by uniform motion around a looped path, instead of a straight path. Then, a clock can be compared with another with which it was previously synchronized, without retracing its path.* Retracing a path reverses the relative velocity and inverts the sign of the vx/c^2 term; this erases the possibility of testing whether such a term is physical or not. However, the coordinates x_1 and x_2, that are considered distinct with the separation $l = x_1 - x_2$ on a straight path, become identical at the same location when the path is looped around. Then, the cat is out of the bag. It is clear why one who applies the STR has to keep chanting the restriction of inertial motion, while most real situations in which it is applied involve non-inertial motion and non-inertial frames. In most of such situations, the prediction of the theory matches the prediction of the 'absolute' theory, like the Lorentz–Poincaré ether theory. But, in those situations where they differ, the experimental result always contradicts Einstein's theory and conforms perfectly to relativity with an absolute privileged frame.

When we examined the motional modifications of the GNSS clocks, the real nature of the $\vec{v} \cdot \vec{x}/c^2$ term was clarified further. Appearing as the clock correction $\vec{D}_s \cdot \vec{v}_r/c^2$, *the term arises right in the rest frame of the observer* (receiver), because of the Galilean nature of the one-way propagation of light, violating Einstein's light hypothesis.

Thus, we have now multiple proofs that the Lorentz local time is not a real physical modification of time; it is merely a mathematical device to make the Maxwell electrodynamics invariant under motion. Apart from the physics of the GNSS clocks, we have the reliable direct experimental proof of the Galilean nature of light, and we have a convincing logical proof of the conflict between Einstein's light hypothesis and his own conclusions of relative simultaneity and clock synchronization in the STR. All these clearly show that both time and simultaneity have the status of the absolute, as supported by the cosmic time and Cosmic Relativity.

7.4 Summary

We were engaged in a thorough analysis of the motional modifications of the atomic clocks of the GNSS and the corrections required for accurate navigation. After accounting for the gravitational time dilation of the satellite clocks relative to the reference clocks in the stations on the Earth, we examined in detail the motional relativistic correction. We found that the factual correction for time dilation is not according to the prescription in the STR. Instead of the relative velocity of the satellite clocks relative to the receiver, the corrections use the individual velocities of both the satellite clocks and the receiver clock, exactly as stipulated by Cosmic Relativity.

Further, there is a correction applied to the GNSS clocks that is first order in the velocity of the receiver, irrespective of whether its motion is inertial or non-inertial, of the form $\vec{D}_s \cdot \vec{v}_r/c^2$. This turned out to be entirely due to the Galilean propagation of light, with the relative velocity $\vec{c} - \vec{v}_r$. We studied multiple avatars of this correction, including the one applicable when the GNSS receiver is at rest on the Earth's surface at various latitudes.

Our journey exploring the time of the moving clocks of the GNSS showed us that the navigational aids in a light smartphone signify the heavy burden of carrying the wrong theory of physical relativity and clocks. Fortunately, we are ready with the correct theory based on the cosmic gravitational paradigm that accurately and completely takes care of the relativistic modifications of the GNSS clocks.

We have come a long way in our understanding of the fundamental as well as the technological aspects of the gravitational and motional modifications of time. The all-important insight gained, with multiple evidence, is the realization that the motional modification is factually the cosmic gravitational time dilation of moving clocks. Hence, there is only one universal kind of modification of time – that is the gravitational time dilation.

Chapter 8

The Dissemination of Standard Time

Human interactions and the management of everything from programs of daily life to affairs of States, depend on the global distribution of a common time. Such a 'standard time' needs to be spatially transferable as well as accurate enough for the intended application. For example, for most human activities, the global synchronization at the level of seconds might be enough, but modern commerce demands the accuracy of a fraction of a second (the typical duration of human reflex reaction!), astronomy needs sub-microsecond synchronization, and satellite-based navigation requires sub-nanosecond accuracy in the common time. For some purposes, the transfer of time could be on a small spatial scale, and for some others the reliable transfer over the entire globe becomes necessary. The maintenance and distribution of a stable common time is an involved technical problem that depends on the advances and understanding of physics, mathematics, and technology.

The standard time is the time kept by a set of standard clocks, which are agreed as the 'reference standards' by the mutual consensus and conviction of an empowered body. A standard clock is not always the most accurate and stable clock available in laboratories; what is important is the ease with which the standard time can be locally accessed, reproduced, and synchronized at

Gravity's Time
C. S. Unnikrishnan

ISBN 978-981-4877-72-5 (Hardcover), 978-1-003-25656-4 (eBook)
www.jennystanford.com

locations where it is required, keeping the foremost criterion that its accuracy and stability are sufficient for all applications envisaged. Thus, *the atomic time based on the microwave transition in the Cesium atom has been used as the standard of time for over half a century now.* The technology of the operation of the Cesium standard has drastically improved, enabling nearly 100-fold improvement in the Cesium atomic clocks. The stability of the current Cesium atomic fountain clocks, based on advances in laser cooling of atoms to micro-Kelvin temperatures, is about 10^{-16}. This means that two such clocks, once synchronized, show the same time within a nanosecond for an entire year. Clearly, there is no urgent need to look for another better standard in the near future for any conceivable application. However, the research in fundamental physics benefits from even better clocks and the best atomic clocks based on the optical transitions in ultra-cold atoms and ions are a further 100 times better than even the impressive Cesium fountain clocks! What is most interesting is that people have found ways of distributing and cross-synchronizing such 'super-clocks' across countries.

Apart from the practical difficulties, which are numerous, one should always keep in mind that both the gravitational time dilation and the motional time dilation of clocks are at work. The latter is the cosmic gravitational time dilation due to any motion relative to the matter distribution in the Universe, as described by Cosmic Relativity. Just to give an idea of the magnitudes, a mere 10 cm of level difference between two optical super-clocks in a terrestrial laboratory amounts to the relative gravitational time dilation of 10^{-17} s/s. The clocks themselves are precise enough to show a second to 17 decimals! Then, the two clocks will measurably drift away from each other, and show different times a few seconds after their synchronization. Also the motional time dilation is measurable even with the low relative velocity of a metre per second.

I will now briefly review how the standard time is accurately distributed globally. 'Review', because we have already seen the possibilities and methods of the one-way and two-way transfer and synchronization of time, as well as the subtle traps of the physics of relativity that one needs to understand. What I would like to highlight is that the one-way transfer of time and its synchronization are still lacking the support of correct physics.

This is because the community of time keepers has not realized that the factual propagation of light, which is used for the transfer and synchronization, is Galilean and that the time dilation of the moving clocks depends on their absolute velocity relative to the cosmic frame, and not merely on their mutual relative velocity. The science and technology of the dissemination of the standard time are obviously incomplete without this factual physical wisdom.

8.1 The Standard Time

The standard of physical time had evolved from astronomical references to atomic references by the 1950s. Before the advent of the atomic clocks, there was only one standard clock, the rotating Earth. Even then, the astronomical observations required to determine the standard time were being coordinated between different observatories. The Universal Time (UT) was based on such coordination of observations of the earth-clock (the UT in common use is referred to as UT1). The standard second of the earth-clock was defined as 1/86400 of the mean solar day (the time it takes for the repeated apparent passage of the Sun). This is of course slightly different from the shorter sidereal day, defined in terms of the repeated passage of a reference star on the celestial sphere. The sidereal day is 240 s shorter than the mean solar day. The difference arises because the Earth is moving in its orbit around the Sun, in the same sense as its rotation, and therefore the Sun has an apparent 'backward' motion of approximately 1 degree per day relative to the distant stars.

The modern standard of time is based on the hyperfine transition in the lowest (ground) energy state of the Cesium atom, with the particular isotope of atomic weight 133 units (different isotopes have different hyperfine transition frequencies). This corresponds to the difference in the energy of the electron in the two possible states, one state with its magnetic moment aligned with the magnetic moment of the nucleus, and the other with the moments oppositely oriented. The transition is effected with microwaves of frequency around 9.192 GHz, and the Cesium frequency standard is the basis for the Cesium atomic clock. But, unlike the rotating Earth, there

Figure 8.1 The core of the Cesium atomic clocks based on a launched 'fountain' of laser cooled atoms of Cesium. Left, the clock at NPL (National Physical Laboratory)-UK. Right, the clock at NPL-India. The vacuum chamber structure in which the bunches of very slow atoms are launched for measurements is shown in the middle picture, along with the scheme of the 'spin-flipping' hyperfine transition at 9.192 GHz, involving the interaction of the magnetic moment of the electron (e) and the smaller moment of the nucleus (N).

is no 'unique' standard atomic clock. Several laboratories construct and operate similar atomic clocks (Fig. 8.1). The standard clock is an average 'virtual clock' based on several such standard clocks using the Cesium transition as the reference. The international atomic time (TAI, from the French designation 'Temps Atomique International') is based on such clocks. Within this scheme itself there are specially constructed primary clock standards, operational in a small number of laboratories, and a large number of secondary commercial atomic clocks that are operational in most countries (McCarthy and Seidelmann, 2009; Jones, 2000). The two are linked by time transfer protocols to ensure corrections in the secondary clocks. Before the invention of the atomic reference for the standard second, the ephemeris time (ET) was being discussed, for a short period, as the stable universal standard for the international time and the SI second (the unit of time in the 'Système International d'unités). It may be recalled that the ET was based on the stable Keplerian orbits of the solar system bodies like the Earth and the Moon. The ET was in fact adopted by the International Astronomical

Union to define the standard second in the SI units, before the ET went obsolete. The ephemeris second (defined as 1/31556925.975 of the tropical year in 1900) is slightly shorter than the second defined by the rotation of the Earth, as 1/86400 of the mean solar day (UT). Hence, two hypothetical clocks that use these two different standard seconds would progressively drift from each other, with the UT clock running behind (reading less time). The difference is very small, equivalent to about 1.7 ms per day, with the UT clock lagging a second behind the ET clock in about 580 days.

The ephemeris standard was quickly replaced by the atomic standard, soon after the invention of the functional Cesium standard. The SI second based on the atomic transition was defined as the time equivalent to 9,192,631,770 cycles of the radiation corresponding to the transition between the two hyperfine levels of the ground state of the Cesium-133 atom. This particular number was chosen because the atomic second was matched to the standard second of the ephemeris time, and not to the 'slower' second based on the rotation of the Earth, which was the basis of the UT, used in all terrestrial applications (matching the atomic second to the solar second would have required another 167 cycles of the 9.192 GHz microwaves, in a second). Therefore, the new international atomic time TAI was destined at birth to progressively diverge away from the UT, at 1.7 ms/day, even if the rotation of the Earth were stable.

The continued slowing down of the rotation of the Earth also contributes to the progressive drift of the UT from the atomic standard time, the TAI. But the rate of slowing is not regular and has many unpredictable features. It turns out that international time-keeping is a complicated affair. The Coordinated Universal Time (UTC) is also based on the standard atomic second now, but adjusted to be close to the factual solar time (UT). Thus, UTC is the 'usable time' based on atomic standards, for all purposes. The TAI is just a stable reference, ticking away in the background. Since the atomic time can be assumed to be stable (fluctuating by less than a nanosecond per day), the UTC clock needs to be adjusted periodically to bring it close to the UT, which is slowly drifting away, behind. This is done by adding 'leap seconds' every few years to the UTC. That is, UTC+ N number of leap seconds = TAI. In effect, the UTC provides the link between the TAI and

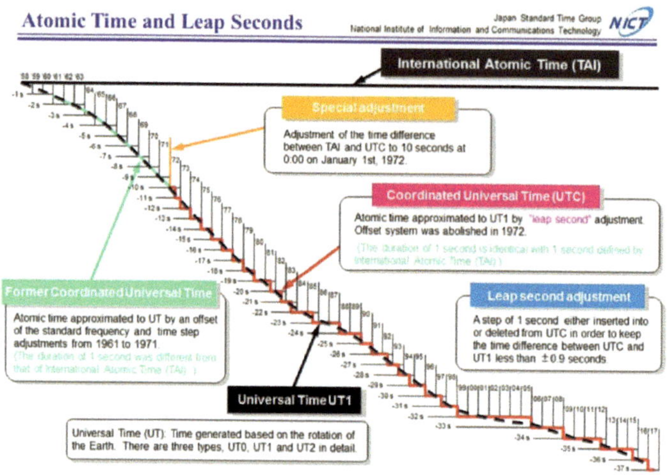

Figure 8.2 The terrestrial universal time UT1 drifts from the atomic time TAI. The coordinated universal time UTC is maintained close to the UT1. Leap seconds are to be added to cover the drifting difference between the UTC and the TAI. The practice started in 1972, when 10 s were added in one step. By 2018, 27 more seconds were added (this self-explanatory compilation is from Japan Standard Time Group, NICT).

the UT. Since there are unpredictable fluctuations in the rotation rate of the Earth, the adjustments between the UTC and the UT needs careful astronomical observations. At the present average rate, the day would be longer by a millisecond in less than 100 years. Therefore, from 1970 to now, the mean solar day has dilated to about 86400.001 s, requiring the UTC to be adjusted by a whole 'leap' second every few years, due to the slowing of the rotation. In fact, the TAI-UTC synchronization started in 1972 with an adjustment of the UTC a whole 10 s behind the TAI, to correct for the 'past drifts', bringing it close to the UT within about 50 ms (Fig. 8.2). During the past 48 years, another 27 leap seconds have been added to the UTC, to compensate the difference of 37 s between a standard atomic clock and the earth-clock, accurate to a fraction of a second. (The reference time in the GPS (GPST) is in between the UTC and the TAI; the GPST was matched to the UTC in 1980 and it is now 18 s behind the UTC because there are no leap seconds in the GPST.) I am sure some experts would be complaining and lamenting that if we had chosen to match the atomic SI second to the mean solar day second, instead of the ephemeris second, the UTC and the TAI would have been still pretty close, within 4 or 5 s (only due to the slowing down of the rotation of the Earth). Then, it would have remained so for a long time!

It is not possible to overstate that we should be thankful as a society to the large number of people maintaining the international time TAI, and its management and dissemination in a human-usable format as the GPST and the UT1. The atomic standard will not see any drastic change anymore, though an optical transition standard that is more accurate and stable than the Cesium hyperfine transition is likely to be internationally adopted soon.

8.2 Distribution and Synchronization of Time

As emphasized on every occasion, time is a local quantity, as local as the clocks, with only a localized existence in space, as the 'present'. Only when the relations between these local times at different locations are established by explicit communication and interaction, there is the global time. Generating a global time from the local times with accuracy is a difficult task because of many hurdles, some of fundamental physics, and some arising in practical considerations.

The transport of clocks is a method for the transfer of time, in principle, but this is not practical. So, we will not discuss it in this section. The relevant physical principles and the importance of the modification of the rate of clocks during such transfer, for the theories of relativity and the measurement of the relative speed of light, are discussed earlier. The only practical methods of time transfer and synchronization are using the electromagnetic signals like radio-waves and light.

The foremost considerations of fundamental physics in such methods are (a) the nature of propagation of the communication signals, (b) the behaviour of moving clocks, and (c) the modification of local times by local physical interactions, especially gravity. We now know from Cosmic Relativity that all the three issues have to do with gravity; the first with the relation between gravity and the propagation of light, and the other two, with the relation between time and gravity.

The first issue concerns the difference between the Galilean propagation that depends on the observer's motion, and the Einsteinean propagation with the invariant relative velocity. With typical terrestrial distances d between standard clocks exceeding 10000 km and the speed of local frames of a few hundred metres per second, the difference between the two scenarios amounts to $d/c \pm v \simeq (d/c)(1 \mp v/c) \simeq 50 - 200$ ns. This is much larger than even the moderate accuracy of 10^{-12} s maintained in the common navigational clocks.

The motional time dilations are of the order $v^2/2c^2$. Since the clocks in comparison are in motion near the Earth or in satellites and other mobiles, the velocity does not exceed 10 km/s. The velocity that is common to the clocks does not figure in the calculation of the *relative* time dilation; they cancel when we consider the relative time dilation. But, the calculations in the STR and in a theory with a master frame like the Universe are different, as we have seen in detail for the clocks of the GNS systems. The STR expression $\delta T/T \simeq -v_r^2/2c^2$ involves only the relative velocity between the clocks, v_r. In contrast, the expression in Cosmic Relativity involves explicitly the speeds of both the clocks in the absolute frame of the cosmic space, $\delta T/T \simeq (-v_1^2/2c^2) - (-v_2^2/2c^2)$. In any case, the corrections needed to maintain accurate and synchronized time are up to a nanosecond per second.

Gravitational time dilations are well characterized because of the accurate knowledge of the Earth's gravitational potential. The mapping of the potential has reached remarkable finesse with cleverly designed precision satellite missions, from the simple LAGEOS satellites to the recent drag-free GOCE project, as well as the twin-satellites GRACE mission (Gruber *et al.*, 2019). The gravitational corrections are $\delta T/T = GM/Rc^2$ for individual clocks and the relative rates range from 0.1 fs/s, for clocks separated by a metre in height on the surface of the Earth, to 0.5 ns/s between the clocks on the Earth and the clocks in navigational satellites.

From the large corrections required, we can get an idea of the immense challenges for maintaining a standardized and stable global time. What is impressive is that this challenge was met directly and there are very good protocols developed. Their only deficiency, a serious one, is that they continue to use the incorrect theory of the propagation of light and the motional time dilation, while managing the required corrections. This is done by making all the relevant calculations in the minimal absolute frame of the two clocks, after subtracting the common velocities, instead of calculating in the rest frame of one of the clocks! In the typical cases, this is the Earth-centric non-rotating frame. In other words, there is no consistency between the actual expressions used (which are of a theory with an absolute frame) and the prevalent physical theory, which is the STR. There is a clear and severe logical gap. As emphasized before, the only logical and empirical choice is to accept the factual gravitational paradigm and the theory of Cosmic Relativity as the basis of relativistic physics.

8.3 Time Transfer with Electromagnetic Waves

A practical way of transferring precise time, adequate for most purposes, was realized already in the beginning of 20th century, right after the demonstrations of the inter-continental transmission of electromagnetic radio signals. The bare round trip duration for such signals around the globe is less than 140 ms. Therefore, knowing and correcting for the approximate distance, it looks practical to synchronize clocks globally accurate to a millisecond using the simple transmission from a radio station. The standard clocks became more precise only much later. With the super stable quartz clocks, and later with the development of the atomic clocks, the situation changed drastically.

Radio communication was demonstrated as a future technology of communication before Einstein's theory of relativity was established. The relevant later input from Einstein's theory was of course the light hypothesis: that the relative velocity of light (radio waves) in any moving frame is the same as it is in a stationary frame. Then, one need not bother about the velocity of the reference frame in estimating the relation between distance and time. But, the light hypothesis is merely a hypothesis. The hypothesis that the relative velocity is Galilean, linearly dependent on the velocity of the observer, is equally consistent, empirically, as we have seen when we analysed several relevant experiments. Indeed, the difference between the two in the context of time transfer and synchronization is very large and readily observable.

8.3.1 Two-Way Transfer

The two-way time transfer had been discussed in detail by Poincaré, in the context of modern relativity. This is also the Einstein synchronization protocol.

The scheme of two-way transfer (TWT) starts by sending a light signal from the clock A at t_{As} to the clock B, separated by an unknown distance L. When B receives the signal, it sets the time as $t_B = 0$, and sends back the signal without delay (or a precisely known delay). When A receives this signal at t_{Ar}, the time at clock A is adjusted to be $t_A = 0 \equiv (t_{As} + t_{Ar})/2$. Then, the clocks are deemed synchronized.

In Einstein's theory, the confidence regarding synchronization is because of the light hypothesis of the STR. Therefore, if the clocks at relative rest are synchronized using light according the TWT, they are assumed to read exactly the same time. However, in Poincaré's view, *the synchronization is assumed to be valid by convention because the relative speed of light is fixed by a convention and not from empirical verification*. The two clocks can be moving together and still be at *relative* rest. Then, if the relative speed of light is factually Galilean, there is a residual excess duration of propagation given by $\delta t = t_B - t_A = Lv/c^2$. However, since v is unmeasurable, this cannot be determined even if the separation L is measured and known. Therefore, the statement of synchronization is a convention, ignoring such a constant shift in the relative time.

Thus, *two clocks that are synchronized at relative rest using the two-way protocol may have a constant synchronization gap that is not detectable*. Hence, such apparent synchronization is consistent, though there is no true synchronization.

8.3.2 One-Way Transfer

This is commonly used in satellite-based systems like the GPS and relies on the absolute faith in the STR and its light hypothesis. As we have seen already in the case of the GNSS, the scheme reveals the inconsistency of using the prevalent theory. Hence, I will analyse it in some detail.

To clearly understand in context the difference between the Galilean absolute view with a variable relative velocity of light and the Einsteinean view with the invariant relative velocity, I will consider the transfer of the standard time between two terrestrial stations at relative rest, say from Paris to Delhi, for the synchronization of the standard Cesium fountain clocks in the national laboratories. The desired synchronization is at the level of 100 fs (10^{-13} s).

The duration of propagation for the distance of 6600 km is 22 ms. The Earth is moving through the Universe, along with the solar system, at about $v_{ss} \simeq 370$ km/s, and at the speed 30 km/s in its orbit around the Sun. The velocity of the stations due to the rotation of the Earth is about 310 m/s (Paris) and 410 m/s (Delhi). Taking the largest of the velocities, the distance travelled in the cosmic space by the Earth is a whopping 8 km during the propagation of the signal between the two stations. Equivalently, the relative velocity of light varies approximately between $c \pm v_{ss}$, in Cosmic Relativity, depending on the phase of the orbital and rotational motion of the Earth.

When a synchronization signal from the station P is received at the station D, the clocks at D is set to zero (Fig. 8.3). The Einstein synchronization of the STR stipulates that a return signal sent back to P can be used for the correct synchronization by setting the clocks at P to $0 = (t_s + t_r)/2 = T/2$, where T is the duration for the two-way propagation. (The Hafele–Keating experiment might have taught a partial lesson to correct for the velocity of the stations due to the terrestrial rotation: the light from P to D is delayed by 25 ns). In general, the Einstein synchronization is simple, having used the light hypothesis. But, that does not do justice to the factual situation! The clocks in Delhi could be a good 27μs shifted from each other because light travels not at the velocity c from P to D, but at a velocity differing by as much as 400 km/s from c. That is a billion times larger gap in synchronization than the intrinsic long-term accuracy of the clocks. But there is no way to detect this difference because a return pulse reaches back to P taking the same duration with the opposite sign; then the Galilean propagation delays cancel and the two-way duration will have no evidence of this one-way aberration.

One may think that if we are patient enough to repeat the synchronization process and monitor the one-way delay, as the Earth rotates to progressively change the direction of 'P to D' relative to the velocity of the solar system, and eventually reverse it, we will see the evidence for the Galilean nature of the propagation of light. If, instead, no such variation is seen, then we have a direct experimental proof of Einstein's light hypothesis, of the invariance of the relative velocity of light. But alas! This is the point on which most people who were concerned with the one-way experiments have erred. The signal in such experiments results from two different physical sources: one is the nature of propagation of light and the other is the time dilation of the clocks due to their real motion. The experimenters who included only one of these to calculate their expected signal, ignoring or forgetting the other in a one-way experiment to 'prove' the invariance of the relative velocity of light, have erred in a fundamental way. The error is the same as the one M. Ruderfer had pointed out in an erratum when he himself made the error in his published paper (Ruderfer, 1960). I mentioned that incident earlier (Section 6.2).

Let us work through the calculation to see this fundamental error, of not understanding the intertwining of the propagation of light between two points and the synchronization of the clocks at these locations. Referring to Fig. 8.3, we consider the experiment to determine the excess delay in the propagation in the direction from P to D, $\vec{L}$, in the rotating terrestrial frame that is also moving through the cosmic frame at the velocity $\vec{v}$. Then the relative velocity of light in the cosmic gravitational paradigm (Cosmic Relativity) is $\vec{c} - \vec{v}$, where $\vec{c}$ is along $\vec{L}$. The extra delay of a synchronizing signal from P to D is then

$$\begin{aligned}\delta t &= \frac{L}{|\vec{c}-\vec{v}|} - \frac{L}{c} \simeq \frac{L}{c(1-\vec{c}\cdot\vec{v}/c^2)} - \frac{L}{c} \simeq \frac{\vec{L}\cdot\vec{v}}{c^2} \\ &= Lv\cos\omega t/c^2\end{aligned} \tag{8.1}$$

As the Earth rotates, the angle between the P to D distance $\vec{L}$ and the global velocity $\vec{v}$ (which remains the same) changes and the product $\vec{L}\cdot\vec{v}$ changes, eventually going to zero as $Lv\cos\omega t$, when

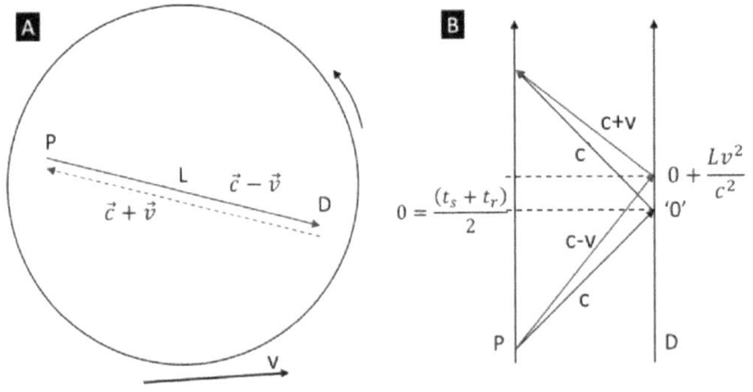

Figure 8.3 (A) Clock comparison and synchronization between two stations using free space electromagnetic signals on the Earth that is moving and rotating in the cosmic space. (B) The belief is that the signal propagates at equal speeds both ways, whereas the truth could be that the propagation is Galilean with the asymmetric relative speeds $c-v$ and $c+v$. Neither the two-way protocol nor the one-way protocol can sense the difference with separated locations, consistent with the principle of relativity.

the rotation completes a quarter turn. The delay reverses in sign and reaches a maximum at half the rotation. Then the expectation is the sinusoidally changing propagation delay, with the peak amplitude of $Lv/c^2 \simeq 27$ μs (this is approximate and indicative, because the actual value depends on the actual direction of v relative to the plane of the Earth's orbit, etc.) This is an easily measurable large signal. The absence of such a signal is what many experimenters relied on to claim as a proof of Einstein's light hypothesis (Krisher, 1980).

However, they missed the crucial fact of the time dilation of the clocks used in the experiment, which are moving in the cosmic frame! We already know, directly from experiments like the Hafele–Keating experiment or the GPS navigation, that two clocks at relative rest in a rotating frame have an extra relative time dilation, in spite of them being at relative rest. In the present case, note that the clocks at P and D are in linear motion in the cosmic space at the velocity $\vec{v}$, while rotating with the Earth. *The clocks are being slowly transported in opposite directions by the rotation of the Earth* over and above the large common velocity they have. Therefore, their common time dilation $\delta t/t = -v^2/2c^2$ is *modulated by the rotation* as $-\vec{v}\delta\vec{v}/c^2$. Then, their relative time dilation (D relative to P) is given by

$$\frac{\Delta t}{t} = -\frac{\Delta t_D}{t} - \frac{\Delta t_P}{t} \simeq \frac{\vec{v}\cdot\vec{v}_r}{c^2} \quad (8.2)$$

Since $tv_r = \delta s$ is the distance by which the clocks are transported due to the rotational velocity, the quantity $\vec{v}\cdot\vec{v}_r t$ is the component of that distance along the gross velocity v. This is

$$\vec{v}\cdot\vec{v}_r t = \vec{v}\cdot\vec{\delta s} = v(L - L\cos\omega t) = vL(1 - \cos\omega t) \quad (8.3)$$

We get the relative time dilation of the clocks,

$$\Delta t = t\frac{\vec{v}\cdot\vec{v}_r}{c^2} = vL(1 - \cos\omega t)/c^2 \quad (8.4)$$

The total time delay is the sum of the excess propagation delay δt (eq. 8.1) and the factual time dilation of the spatially separated clocks Δt (eq. 8.4),

$$\delta t + \Delta t = \frac{Lv\cos\omega t}{c^2} + \frac{vL(1 - \cos\omega t)}{c^2} = \frac{Lv}{c^2} \quad (8.5)$$

This expression with a constant unmodulated signal (null signal) is a remarkable relation, signifying the principle of relativity, while highlighting the Galilean nature of the propagation of light. The experiment will be unable to see any change in the signal in spite of the fact that the propagation delay is changing due to the variable relative velocity of light, because the relative time dilation of the clocks is changing in the opposite sense, *effacing the oscillatory evidence*. This is the reason why one-way experiments that use spatially separated clocks are incapable of revealing the real nature of the propagation of light. What I have been discussing is the general case of the situation that was pointed out in the self-correcting 'erratum' published by M. Ruderfer half a century ago. All past one-way experiments have ended up with the incorrect conclusion, because of not realizing this most fundamental feature of relativity.

However, let us think of the situation carefully. We all know that the Universe has matter and its average density is measured rather accurately. We know how to estimate the magnitude of the gravitational potential of the cosmic matter, at least approximately. This necessarily implies the gravitational time dilation of clocks moving relative to the cosmic matter. At least one has to accept that the clocks in rotating frames have an anomalous time dilation, as demonstrated in the Hafele–Keating experiment or the GPS. Hence, one has to accept that our expression for ΔT is more or less correct. This implies that a null signal in the experiments that monitor for the variation in the propagation delay is a clear evidence for the Galilean nature of the propagation of light, of the dependence of the relative velocity of light on the velocity of the observer. If the velocity of light were an invariant constant, the uncompensated time dilation during the rotation would have stood out as the signal in the experiments. *A null signal is in fact the proof for the mutual cancellation*. The two go together, and it is easy to understand. The larger the velocity of the frame, the lower is the relative velocity of light $(c - v)$, and the larger is the propagation delay. Then, one would expect a larger duration of propagation between the times marked by clocks, $+\delta t$. But, the larger the velocity, the larger is the relative time dilation of the clocks, $-\Delta t$. That is how we end up with the situation of both the effects remaining unobservable, the two physical effects having intertwined with opposite signs.

The conclusion is that once we acknowledge the reality of the matter in the Universe, such one-way experiments reveal definite proofs for the Galilean nature of the propagation of light; their null signal is the clear evidence that falsifies the light hypothesis and the Special Theory of Relativity. The clocks at two locations, when 'synchronized' using electromagnetic signals, are really running with a synchronization gap that depends on the distance and velocity of the frame in which the clocks are at relative rest. However, this gap is not measurable because a comparison of the clocks requires either a signal in the reverse direction or the physical movement of the clocks to the same location, both of which null the synchronization gap.

8.4 Optical Fibre Links

The transfer of the standard time through dedicated or commercial optical fiber links between stations is a relatively new technology that has achieved a superior performance than the satellite-based time transfer between terrestrial stations. The distance over which the reliable transfer and clock comparisons are done using optical fiber links now exceeds well over a 1000 km. The key advance that made this possible is the availability of very stable lasers at the telecom wavelength of 1.55 μm, as well as the ability to cancel and correct fluctuations in the optical signal by a two-way compensation scheme, where the laser signals pass through the same fiber and optical circuits back and forth between the stations. This has enabled the determination of the optical phase of the carrier wave (wavelength 1.55 μm and frequency 194 THz) with an accuracy of a part in 10^{18}. That is, the optical frequency (2×10^{14} Hz) of a stable laser at one station can be reproduced at another station at the other end of the fiber link, thousands of kilometres away, with the high fidelity equalling the input frequency to a fraction of a milli-Hertz. This is remarkable, considering that there are many sources of perturbations as light passes through the glass fiber link that is buried shallow. This technological feat was possible because of the near-complete cancellations of the perturbations in the *two-way protocol* (Lopez et al., 2015).

The transfer of time over such a link is demonstrated with an impressive accuracy of about 10 ps. This is better than the satellite-based time transfer. What one needs to keep in mind is that the same subtleties of relativity, linked to the Galilean propagation of light and the time dilation of the clocks that depend on their absolute velocity in the cosmic frame, are applicable in the case of the optical fiber links as well. Then, a critical question should naturally arise. We have seen that one important reason why people do not realize that the relative velocity of light depends on the velocity of the observer, is that the time dilation of clocks is intertwined with the propagation delay of light. So far our discussion was about the *propagation of light in free space*, at its bare speed c. However, in both optical fiber links and in the experiments of the one-way propagation of light using an optical fiber, the consideration of relative velocity is very different. Also, the velocity of propagation in the optical fiber is significantly smaller, about $2c/3$ or 2×10^8 m/s (the speed of light through the glass fiber is c/n, where n is the refractive index). Then the propagation delay is correspondingly larger. At the same time, once light is injected into the glass fiber, it is no more propagating freely; there is a definite medium to which the propagation is anchored, very unlike in the empty space. Then the light hypothesis of the STR is not valid in any straightforward manner. If the piece of glass is moved, the light inside should move with it. Shouldn't the effects of the observer-dependent relative velocity be very different now? More importantly, how can the details of the duration of propagation, which are entirely different compared to the situation in free space, match anymore the factual time dilation of clocks that remains the same?! Surely, this is an important question. The very reason for mentioning the topic of the time transfer using the optical fiber links was to pose and answer this question, which gives an additional convincing evidence for the robustness of Cosmic Relativity.

Cosmic Relativity has to answer how the propagation delay through the fiber could remain undetectable, exactly compensating the variable relative time dilation between the clocks that we discussed in the previous section, in spite of the very different speed of propagation. We saw that the relative time dilation oscillates between $\pm Lv/c^2$. We understand how the propagation delay has the same behaviour with the opposite sign, if light is propagating in free space. Naively, one would think that the light injected in the fiber is carried at the speed at which the fiber is moving in the cosmic space. Thus, instead of the relative speed $c - v$ in empty space, one would expect just c/n, relative to an observer who is moving with the fiber, in the laboratory on the Earth. Then, the propagation delay would be a constant, whereas the relative time dilation is variable. The mismatch would manifest as a measurable signal. However, there is a well-known physical phenomenon, first experimentally demonstrated in the 19th century by H. Fizeau in Paris, called the Fresnel drag, after A. Fresnel. I do not go into the details, but the experimental observation was that the light propagating in a medium that itself is moving at the velocity $\pm v$ has the nett velocity $(c/n) \pm (v - v/n^2)$, instead of the naively expected $(c/n) \pm v$.

Cosmic Relativity recognizes the Fresnel drag as due to the 'competition' between the gravitational effect on the propagating light due to the cosmic matter that is flowing past, relative to an observer moving with the Earth in the cosmic space, and the electromagnetic interaction that holds light to the medium of the glass fiber. If the cosmic matter and its gravity were absent, light would have propagated at the relative speed c/n. But the gravitational drag in the opposite direction makes the relative speed velocity dependent, $(c/n) - v/n^2$. Note that the denominator is essentially the strength of the electromagnetic interaction in glass, $n^2 = \varepsilon/\varepsilon_0$, and the numerator is indeed the cosmic gravitomagnetic potential, $-v = -v\Phi_u/c^2$. This matches the Fresnel drag, derived by Fresnel in 1821 from the ether theory in the pre-relativity days. With the relative velocity as $(c/n)-v/n^2$, the amplitude of the excess propagation delay is

$$\delta T = \frac{L}{(c/n) - v/n^2} - \frac{L}{c/n} = \frac{L - L(1 - v/cn)}{(c/n)(1 - v/cn)} \simeq \frac{Lv}{c^2} \tag{8.6}$$

We get the excess propagation delay that is required to exactly match the relative time dilation of two clocks separated by the distance L, the same delay as in free space. All traces of the fact that light was propagating in a medium with the refractive index n have vanished. This is indeed a remarkable success of Cosmic Relativity. In fact, this is also the reason why a fiber-based rotation sensor, relying on the Sagnac effect, works identically to a sensor based on the free space propagation of light (Fig. 8.4C). (The small second-order v^2/c^2 correction also gets cancelled if we include the second-order length

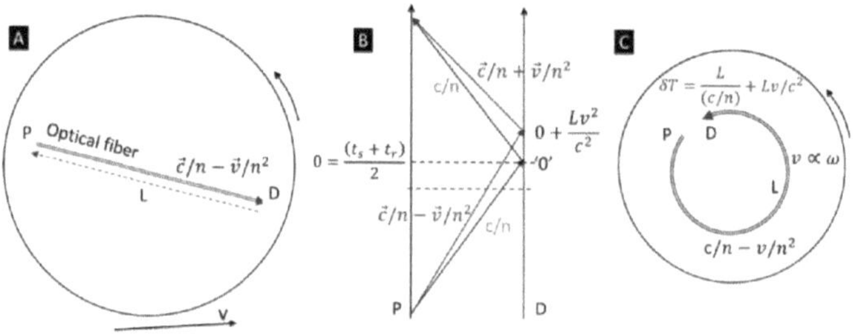

Figure 8.4 Time transfer and synchronization using optical fibers (compare with Fig. 8.2). (A) light propagates through a fiber link between P and D. The relative speed is believed to be c/n, but factually it is $(\vec{c}/n) - \vec{v}/n^2$. (B) The signals in a space-time diagram. The reality is very different from the belief. While the two match at the end of the two-way protocol, the signal is slower one-way and reaches the destination later than what is believed. The return is correspondingly faster. (C) The truth becomes observable if the stations P and D are nearby, with an optical fiber link in a rotating frame. The duration an optical pulse takes to come around is not $L/(c/n)$, but the longer $\delta T = L/(c/n) + Lv/c^2$.

contraction of the fiber, but that is a small detail in the context of the one-way experiments.)

8.5 Summary

In this chapter, we saw a brief overview of the standardization of time and its dissemination, by international cooperation. The global synchronization of time and its management are admittedly a complicated task. However, the scheme is now mature, except for the use of the experimentally invalidated and logically unsupported light hypothesis of the STR as the basis for the time transfer and synchronization. Some of the consequences were examined with precision, when we discussed the GNSS earlier.

I analysed in some detail the intertwined fundamental aspects between the propagation of light and the relative time dilation of spatially separated clocks. In particular, I showed that the excess propagation delay, due to the variable relative velocity of the light signals in the one-way transfer of time, matches always exactly the relative time dilation of the clocks at the two stations, rendering the direct detection of the Galilean nature of the propagation impossible. Thus, a large number of experiments that claimed in the past to test and demarcate theories using the one-way propagation of light were shown to be ineffective and invalid.

The time transfer and the comparison of time through optical fiber links have achieved a remarkable and truly impressive performance of late. I mentioned new physics issues that become relevant when the propagation of the messenger waves is in a medium like glass fiber. I showed how the cosmic gravitational drag on the propagation of light in media like glass, equal in character to the well-known Fresnel drag, results in the same behaviour of the excess propagation delay as in free space, matching again the relative time dilation of separated clocks. This was a striking demonstration of the integrity and correctness of the theory of Cosmic Relativity.

We live in a world where humanity no longer shares the resources of the Earth in any fair manner; not even its common resources like air or water, let alone the commercial exploits. However, we do share a 'common time' and its technology, seemingly quite fairly, with a true international participation. What is missing from a satisfactory conclusion on this is the solid foundation of the correct theory of relativistic time. Is it too much to expect of us to acknowledge the encompassing gravitational role of the entire Universe in the physics we share, when we are hesitant to acknowledge the reality of a common Earth? I hope not.

Chapter 9

Gravity's Time in Quantum Mechanics

Quantum mechanics (QM) is the description of dynamics in the microscopic world of atoms, photons, electrons, etc. Similar to Newton's law of dynamics, $dp/dt = F$, which is the statement on the time evolution of the dynamical physical state associated with matter, the Schrödinger equation $i\hbar d\psi/dt = H\psi$ is the statement of the time evolution on the quantum physical state of matter, denoted with the function $\psi(x, t)$. Though the wavefunction $\psi(x, t)$ is not a dynamical quantity like momentum, the role of time in the description of dynamics is similar and straightforward, if the final results are interpreted as the average over a statistical ensemble (collection). In the conventional interpretation of quantum mechanics, the wavefunction is supposed to be the complete description of the physical state of the particle, like an electron or an atom. It is complex-valued, mathematically, and needs two real quantities for its specification. It evolves in 'magnitude' and 'phase' as the reference clock advances in time. The evolution progresses causally, connecting the past to the present and into the future, as one expects in dynamics. Thus, the variable called 'time' in the microscopic world of quantum mechanics is the same as the time that is familiar in our macroscopic world.

Gravity's Time
C. S. Unnikrishnan

ISBN 978-981-4877-72-5 (Hardcover), 978-1-003-25656-4 (eBook)
www.jennystanford.com

The time variable of QM is provided by physical clocks, just as in macroscopic classical mechanics. But, such macroscopic clocks are made out of microscopic atoms, which obey quantum dynamics. This immediately brings the realization that the stable time marked by any clock is an 'average' time, where the individual atoms behave slightly differently from each other, causing a small 'quantum uncertainty' around the mean time. This is particularly clear in the case of an atomic clock, where each Cesium atom has a frequency that is well defined, yet different from other Cesium atoms by a tiny amount dictated by the Heisenberg uncertainty relation. The quantum uncertainty limits the ultimate stability and accuracy of a clock. This means that two identically constructed atomic clocks are expected to drift away from each other after synchronization, because of the small difference in their exact rate. Yet, each is as good as any other, marking legitimate time. Similarly, for atomic clocks based on a single ion, the frequency of the individual photons that probe the transition varies slightly, and the average of a large number of photons at resonance results in the final accuracy attained. Therefore, the standard of time that we use now is really that of a virtual clock, generated by averaging the time marked by several best atomic clocks around the globe.

In all these examples, the central dictum that time is nothing but the evolution of a physical system is valid. We are, in all cases conceivable, comparing the evolution of one system with that of another, which is chosen as a standard uniform evolution, and hence called time.

Though there are many interesting topics related to the presence and role of time in quantum mechanics, we will discuss only a small subset in this brief chapter; our focus will be the topic of quantum correlations and the issues of causality and locality of spatially separated events. The specific context is the absolute universal time available in the evolving Universe. The relation between the results of measurements (observations) at two locations is a 'correlation'. In particular, I will discuss the question of simultaneity and time-order of quantum measurements at different spatial locations. This is a topic that has been debated since the days when Einstein pointed out a fundamental issue that questioned the theoretical completeness of quantum mechanics as an encompassing theory. The notion of causality – a well-defined order between a physical cause and its effect – relies on a well-defined and invariant order of events in time. Causes happen before effects, by definition. It should not happen that a phenomenon that was assigned the status of the cause of an effect by some witnessing observers, occurs for some others, after the effect. Such a divide between individual observers would result in a world that would be physically inconsistent and incoherent, and psychologically disastrous.

Closely related is the notion of locality in physics. A physical event in one spatial location should be able to influence any phenomenon at another location only after a delay, dictated by the fundamental constraints of physics, like a maximum possible velocity of propagation of any kind of influence in this Universe. Telepathy that violates such constraints is deemed an impossibility in the physical world.

The study of the gravity of the matter in the Universe showed us that the Universe is the master frame for the dynamics of every physical entity. Cosmic gravity limits the speed of propagation of every form of energy and matter, be it light, gravitational waves, or any other physical entity. The evolving cosmos defines the reference for motion and it provides a universal time, which is globally synchronized. Then, it also provides the reference for an absolute and invariant ordered relation between every pair of events; *cosmic gravity makes the relation between a genuine cause and its effect universal, the same for everybody in this Universe, irrespective of their motion and environment.* These are the facts we are sure of now, based on logic, observations, and experimental results. Every propagating physical influence is energy in motion and it is subject to the gravitational interaction. The essential conclusion is that the gravity of the Universe does not allow any kind of physical influence to propagate between spatially separate locations faster than the 'absolute' speed $\sqrt{\Phi_u} = c$.

However, the current beliefs in theoretical quantum physics are in drastic conflict with the rational notions of causality and locality dictated by relativistic physics! This happened because quantum mechanics is a theory centred on a theoretical ingredient called the 'wavefunction' for which an acceptable interpretation is not yet found, even after a century of success in the accurate calculations using that theory. Essentially *most physicists today believe and profess that an observation done at one location can and does instantaneously affect the result of another at a location far away*, even arbitrarily far, from the location of the first observation! Nobody has any idea as to what the agent or mechanism of the super-physical influence is, but the prevalent belief is that such a nonlocal influence is at work in the correlations of physical quantities of two (or more) particles. We will of course see the details of this situation in the subsequent sections. For some unjustified reason, physicists believe that this telepathic influence across locations happens only in the microscopic world, with no leakage into the macroscopic world. If physicists, who understand and impose the 'constraint of locality' on all phenomena, themselves argue for a blatant violation of locality to preserve their faith in a particular understanding of one theory, then it is the sign of the collapse of rationality in theoretical physics. It is ironical that by believing in the violation of locality, physicists are inadvertently subscribing to a world of physics that does not respect any theory of relativity, including Einstein's STR.

While commenting on the situation regarding the general attitude of physicists towards locality in quantum phenomena, Einstein (sarcastically) wrote in the 'Schilpp volume' (Einstein, 1949),

> I close these expositions, which have grown rather lengthy, concerning the interpretation of quantum theory with the reproduction of a brief conversation which I had with an important theoretical physicist. He: "I am inclined to believe in telepathy." I: "This has probably more to do with physics than with psychology." He: "Yes."

This was written before a class of classical statistical theories, called the Local Hidden Variable Theories (LHVT), became prominent in discussions of alternatives to the quantum theory. (Unfortunately, many think that Einstein was an advocate for such theories to replace quantum mechanics though, in reality, Einstein was totally dismissive of any attempt to replace QM with a classical statistical theory.) When the utter failure of the LHVT became obvious in the late seventies, the belief in nonlocality became widespread, ironically, as if the failure can be cured by replacing the 'L' (local) with 'NL' (nonlocal). Nonlocality is in direct conflict with the causal order of physical events in time. I will show that the impression of the violation of locality, and the general belief in the telepathic instantaneous transmission of causal influences in the microscopic physical world, stem from the severe misunderstanding of the theoretical developments that followed a paper written by Einstein and his colleagues in 1935. The concepts of time and simultaneity in physics are universal, applicable identically in both the macroworld and the microworld; the difference is the need to consider statistical averages in the microscopic world, due to its intrinsic indeterminism (Unnikrishnan, 2018b). The cosmic gravitational paradigm provides a reliable and robust structure for rational physics with its familiar notions of locality and causality intact.

9.1 Time, Locality, and Causality

We have already discussed the strictly local nature of the notion of time, existing physically only 'here', and 'now'. This is because time is the physical evolution of matter. The impression of a global time, if any, is only because of the similarity and congruence in the progression of numerous local times. As emphasized earlier, *time resides locally in matter and its global causal connection suitable for comparison happens only through interactions between matter at different locations.*

Local times at two spatial locations can be linked only through a physical link, be it an interaction or a material messenger like light. Once this is clearly stated, the notion of 'instantaneous' is not physically reasonable. For, if instantaneous changes are possible even within the same elementary piece of matter, the sequential nature of changes of physical states will all collapse, with an infinite number of changes happening in zero duration. There cannot be a consistent notion of time in such a world. Given this obvious fact, the notion of locality becomes one of the pillars of physics. Physical events at separated spatial locations can be causally connected only with a distance-dependent temporal delay, measurable with a pair of clocks. This is consistent with the fact that all interactions have finite speeds, and causality is established over spatially separated locations only through the interactions. This structure shows the intimate relation between time and locality. If locality is violated, it should clash with the notion of physical time in one way or another, making physics inconsistent. This is because the violation of locality would imply the existence of an interaction that propagates instantaneously between spatially separated pieces of matter. Such an interaction will not be consistent with the physics of relativity, irrespective of which theory of relativity one is considering.

Causality concerns the order of events (elementary physical phenomena) in time. What is the relation between time and causality? Their relation is necessarily through physical interactions. Changes in the physical states of matter are 'events'. In a single physical system, changes are sequential events and there is an intrinsic sense of ordering. This ordering is synonymous with just counting the events locally. The inevitable sequential nature, because there is only one material system at one location, translates to the monotonous count of the changes of the physical state. Though the change of state from one to another in a piece of elementary matter need not necessarily be called a relation of cause and effect, for generality we will consider the sequence as a cause-effect pair since the new state emerged out of the old state. *Such a transformation is causal, and local.* However, the situation in comparing physical events at separated locations is very different. Then, any invariant ordering across locations makes physical sense only when one event in one location is the genuine cause of another event at another location. This kind of cause-effect relation between events can happen only through a physical interaction. Therefore, the temporal order of such events has an invariant meaning. Note that we are only talking about the order, as preceding and succeeding, and not about the exact duration between the cause-event and the effect-event, for we know that durations are affected by gravity and motion. One may then say that events with a causal relation are what completes the notion of global time. Mere 'noninteracting' events do not suggest time. Causality is what gives life and direction to time. The causal order of events is what we call

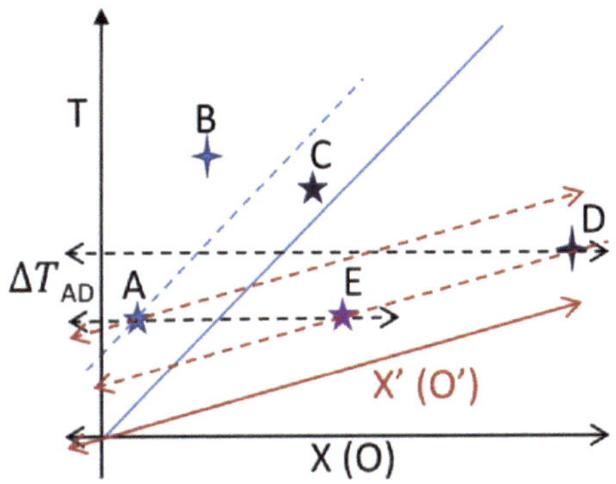

Figure 9.1 A space-time diagram that illustrates the causality relations in the STR. The propagation of light is represented by the 45° lines (blue). For the observer O, the events A and E are simultaneous, and then D, C, and B occur. A and B could be causally related, because B happens after light reaches from A to the location of B, but A cannot be the cause of the other events. For another observer O', the spatial relations (x'-axis) and temporal order are different: E and D are simultaneous, then A, C, and B occur. E can be the cause of C and B, but not of A and D. Importantly, the temporal order of A and D has reversed for O': D precedes A.

the direction of time. That is invariant, the same for all observers. That is how time becomes an integral part of the laws of dynamics.

9.1.1 Causality in Einstein's STR

The order of two events at spatially separated points depends on the state of motion of the observer, in Einstein's STR. This is illustrated in Fig. 9.1.

The observer O records several events A, B, C, D, and E that occur at different spatial locations. This can be represented in a space-time diagram, as shown. The time axis of 'O' is vertical, and all events on a horizontal line (same time) are simultaneous. The propagation of light is represented by the 45° 'light lines'. Any physical propagation from an event to another will be within the wedge between the time axis and the light line from that event. For O, the events A and E happen simultaneously, as per the synchronized clocks in the rest frame of O. Afterwards, the events D, C, and B occur, in that order. There could be a causal relation between events A and B, because B is within the light line from A; A could be the cause and B the effect, but not vice versa. However, there cannot be any causal relation between the event A and C, or A and D, because they are outside the light line from A and cannot be connected with even light, the fastest messenger possible. Another observer O' who is moving relative to O, has a different reference system for the measurement of distances and durations. As we have seen, the separation and the duration are modified by the second-order effects of length contraction and time dilation. However, those are not the ones that affect the ordering in time directly; there is a first-order (v/c) difference due to the relativity of simultaneity that can change the order of the events for the observer O' in the STR. We discussed this in detail in Chapter 6. For O', events are simultaneous when they are in a line parallel to the x'-axis of O'. Thus, events E and D occur simultaneously for O' (the time axis T' is not shown, but the spatial axis, a red dotted line in Fig. 9.1, passes through both events, signifying simultaneous events). Then the events A, C, and B occur. The important point is that for O' the event D occurred before the event A; the order of these events is reversed relative to O.

In contrast, Cosmic Relativity has the universal time, synchronized everywhere, as we have discussed earlier (Sections 4.5 and 4.7). This universal time is also practically manifest in the cosmological markers, like the temperature of the CMBR. Also there is absolute simultaneity. Therefore, there is an invariant meaning to the order of all events, irrespective of whether they are causally related or not. So, according to Cosmic Relativity, if two events, like A and D, have a time order in which A happened before D for some observer, the same time ordering is preserved for all observers. The temporal order of physical events is absolute and objective in Cosmic Relativity, independent of the motion of the observers and their reference frames.

The order of two events that are causally connected is invariant in both the STR as well as Cosmic Relativity, as expected. In both theories, no causal influence can connect two events instantaneously, since the speed of the fastest influence is limited to the speed of light c in the cosmic frame in Cosmic Relativity, and in any inertial frame in the STR. However, it is clear from the discussion so far that if one links two events like A and D by a causal relation – say, A as the cause of the outcome D – it has no invariant meaning in a theory like the STR. For another observer who is relatively moving, A occurs after D and hence, A couldn't be the cause of D.

Summarizing, the violation of locality – causally linking two spatially separated events with an influence that violates the fundamental restriction of the theory of relativity (whether Cosmic Relativity or the STR) – is an unphysical proposal. When stated in the context of the STR, the stipulation is called 'Einstein locality'. What remains to be seen is whether there is any evidence for the violation of locality, or 'nonlocality'.

9.2 Quantum Measurements and Time

The observation of a physical system to obtain a measure of the physical quantities associated with a material system (consisting of matter and its energy) is fundamental to all physics, both for practical reasons and for constructing accurate theories and models. The underlying assumption, which is reasonable from experience, is that the material body possesses characteristic intrinsic measures like mass, charge, etc., as well as definite values of dynamical variables like energy, momentum, spin, etc., at all times, all of which we can observe and measure with a suitable apparatus. We believe that such physical properties exist, with definite values, whether or not we make an observation on a particular physical system; hence, what we see during a measurement are the properties and dynamical variables the physical system possesses. That is the requirement of 'objectivity' of physical properties, a fundamental tenet of physics. If we do not disturb the system much with our measurement, the system will continue in its dynamics, with a negligible change of its course.

All such expectations of experience are shattered when we deal with the atomic world. That our observations, even if delicately done, can perturb a microscopic system significantly is understandable. But, that is not the sense in which observations in quantum mechanics upset our world view. Apparently, there are many situations where the physical systems do not possess any definite physical property before we observe them! The definite value seems to be materialized in the process of measurement, and nobody has figured out how. This has bizarre consequences.

9.3 Quantum Measurements and the Collapse of the Wavefunction

Quantum mechanics is a peculiar theory of the dynamics. Though the theory is about the dynamics of matter, like electrons, atoms, etc., the Schrödinger equation does not describe the evolution of any physical property of the material particle! This may be contrasted with Newton's law of motion (dynamics), which describes how the momentum of a body changes when a force is applied, $dp/dt = F$. The Schrödinger equation $\partial\psi/\partial t \propto H\psi$, however, is for the evolution of an abstract entity called the 'wavefunction', or more appropriately a 'ψ-function. Here, H is a function of the momentum and position, called the 'Hamiltonian', named after W. R. Hamilton. There is a similar looking, but different, description of Newtonian dynamics as well, symbolically with the same Hamiltonian function, written as $\partial S/\partial t = -H$. This was formulated by Hamilton in 1834. Here, S is called the action, which is well defined in terms of the dynamical variables like the momentum, and the coordinates like position, etc. But, *nobody knows* what the ψ-function of quantum mechanics means! This does not diminish the usefulness of quantum mechanics as a theory to calculate and predict physical effects and their quantitative content accurately, albeit only as statistical averages. The ψ-function can be considered to be a 'bag of information', which releases a value for any chosen observable when probed with the correct tool (called an operator), designed for that particular observable. Thus, the ψ-function is not any of the dynamical variables, like energy, momentum, etc., but the ψ-bag holds the information on all such variables. In fact, the ψ-function is 'dimensionless', unlike the common physical quantities. The ψ-function can be found in every physical situation, and from that the probabilities for the physical system to possess various values of energy, etc., can be calculated readily. In fact, the success of QM as a practical theory is astonishing. But the theory has certain serious birth defects.

When an observation is made on any single copy of a physical system, a result for the relevant physical variable is obtained, as expected. But, if the measurement is repeated on an identically prepared copy, the result could be randomly different. Thus, only the statistical average of the ensemble of results is definite, which the theory accurately predicts. The results of quantum measurements (observations) show stochasticity or fluctuations. This is intrinsic to every quantum phenomenon and it cannot be avoided.

The theory handles this stochasticity by describing the physical state on which measurements are done as a 'superposition' of other physical states. Symbolically, it is written as a linear sum of elementary ψ-functions of the same kind, making up a single resultant ψ-function.

$$|\Psi\rangle = a\,|\psi_a\rangle + b\,|\psi_b\rangle + c\,|\psi_c\rangle \ldots \tag{9.1}$$

Each elementary ψ-function represents a definite state with a definite value for the physical quantity like the energy. Then, as in rolling a die or tossing a coin, each measurement randomly picks one of the states in the superposition, say, $|\psi_b\rangle$ for some unknown reason; thus, each measurement has a different result, randomly occurring. The theory has the crucial rule that the probability to get a result is equal to the square of the coefficient $a, b, c, \ldots,$ etc. Since the result is known with certainty, with probability 1 after the measurement is made, all except one of the members of the superposition should vanish instantaneously! If $|\psi_b\rangle$ was realized in a particular measurement, then the whole ψ function reduces to just $|\psi_b\rangle$, with $b = 1$, because it is already 'observed'. This implies that the other parts, $|\psi_a\rangle$, $|\psi_c\rangle$, etc., disappeared instantaneously, $a, c, \ldots = 0$. If the ψ-function has any physical reality, in the ontological sense, this sudden disappearance from the extended regions of space would be bizarre. This has the sinister yet evocative name, 'the collapse of the wavefunction'.

Though all this is beyond a coherent description in space and time, as we used to do for the motion of a particle, physicists were happy that they could calculate and predict, and even understand, a large number of important phenomena, using quantum mechanics. After all, the theory showed the way for all the discoveries that changed the technology basis of life in the past one century – electronics, lasers, superconductors, modern chemistry, and molecular biology.

But, the collapse of the wavefunction made the theory very unsatisfactory. First of all, it implied that the physical system did not possess, before the measurement, the definite value that materialized during the measurement. Even more seriously, the theory could not say anything about the dynamics of how a result materialized. There seemed to be an abrupt jump to one of the values, whereas the theory and its equation of smooth evolution do not allow any such jump.

Without elaborating on the many troublesome aspects of the 'problem of measurement' in quantum mechanics, let us go directly to the topic we want to focus on – that of the violation of locality and causality. The problem of collapse becomes impossible to comprehend when there are two (or more) particles or physical systems. This was pointed out by Einstein, and expressed first in a paper by him with two other associates, famously known as the Einstein–Podolsky–Rosen (EPR) incompleteness argument or the 'EPR paradox' (Einstein *et al.*, 1935).

9.4 The EPR Argument of the Incompleteness of QM

The EPR argument relied on the peculiarities of observations on a physical system consisting of two particles, which had an interaction in the past and subsequently separated into two spatial locations, no more interacting. Here we mean interaction in the usual sense in physics, like electromagnetism or gravity. The interaction could also be one by which a particle breaks up into two particles, which then separate. We look at the cases where the interaction resulted in a correlation between the dynamics of the two particles.

Let us first consider two particles in the familiar classical mechanics. If the pair of particles came from the breaking up of one parent particle at zero momentum, the two individual momenta are correlated as $p_1 + p_2 = p_0 = 0$, because of the conservation of the total momentum. *Since we know that the total momentum is zero, if we measure the momentum of one particle we would know the momentum of the other, from simple deduction, without any actual observation on the other.* However, the dynamics of the two particles are independent. The measurement of momentum on one particle enables the deduction of the momentum of the other particle, using the prior knowledge on the correlation $p_0 = 0$. But nobody would claim a cause-effect relation between one measurement and the other, done independently on the two particles that are spatially separated. One measurement is not the cause of the result that could be observed on the other. Since there is no cause-effect relation, there is no privileged order of measurement as well; the observation on one particle could be done without bothering about the observation on the other, at any time, and later compared to confirm the correlation.

How is the same physical problem described in quantum mechanics? In the corresponding quantum mechanical problem, the considerations are entirely different. The QM description is not in terms of the properties of particles, but it relies on a ψ-function. In this case, the QM description is in terms of the joint ψ-function for the two particles, constructed from the ψ-function of each, by making a product that is consistent with the prior interaction and the correlation.

First we will see what happens when there is no interaction or correlation. Earlier we saw that the ψ-function of each particle can be a superposition of ψ-functions representing definite values of dynamical variables. Let us consider the case of just two possible values in our measurement, '+' and '−'. The ψ-function of such a particle would be

$$|\psi\rangle = a\,|\psi_+\rangle + b\,|\psi_-\rangle \tag{9.2}$$

When there are two independent particles A and B, we have two similar ψ-functions ψ_A and ψ_B, and usually the joint ψ-function that represents the two particles together, is expected to look like the general superposition

$$\Psi = \psi_A\psi_B \sim \psi_{A+}\psi_{B+} + \psi_{A+}\psi_{B-} + \psi_{A-}\psi_{B+}\psi_{A-}\psi_{B-} \tag{9.3}$$

This is the superposition of the 'double ψ-functions' representing all possibilities of the results (four, in this example, $++$, $+-$, $-+$ and $--$) of double measurements, one measurement on each particle, done independently. I have omitted the coefficients a,b,..., etc., that determine the probabilities, for simplicity of the discussion. When a measurement is made, say on the particle A, if the outcome happens to be '$+$', then the two possibilities represented by the last two terms that contain the result of A being '$-$' are eliminated and the joint ψ-function collapses to

$$\Psi' = \psi_A \psi_B \sim \psi_{A+}\psi_{B+} + \psi_{A+}\psi_{B-} = \psi_{A+}(\psi_{B+} + \psi_{B-}) \qquad (9.4)$$

The particle A has the definite state '$+$' and B is still in a superposition, to be resolved by an actual measurement. This is still a product of a ψ-function for A and another for B, with two separate factors. In other words, *the joint ψ-function is separable into two individual ψ-functions.* All this is reasonable, given the peculiarity of collapse in QM.

However, when the particle came from a breaking up of a single system for which the conserved physical property had the value 0, then if one particle is in the '$+$' state, *the other has to be in the state* '$-$' (the results $+-$ or $-+$). We cannot have the combinations $++$ and $--$, because that would be inconsistent with the parent state 0, which allows only $+-$ or $-+$. How do we describe this with a ψ-function, the only description we know in QM? That looks easy; the joint state should be something like

$$\Psi \sim \psi_{A+}\psi_{B-} + \psi_{A-}\psi_{B+} \qquad (9.5)$$

This looks as simple as the previous Ψ' of eq. 9.4. However, a disaster for reasonable physics is lurking within! Try as one may, this cannot be written as a product of individual ψ-functions of A and B, $\psi_A\psi_B$. *There is no 'common factor' representing one system to be factored out of the whole, to write it as two separate factors; the 'separability' is lost.* If one asks the question, 'what is the state of A (or B)', the answer is 'it is not $+$, it is not $-$, and it is not a superposition of $+$ and $-$'. But, these are only physical states in the entire theory of QM! Therefore, suddenly, neither particle has any physical state whatsoever, in the otherwise successful theory of QM! What have we done in our pursuit of understanding nature? *We ended up with a theory in which a physical system, like a particle that we can sense with our apparatus, has no physical state whatsoever before an actual measurement.* (This was clearly pointed out first by Schrödinger in 1935, and he called it 'entanglement'.) However, despite this disturbing situation, let us go further.

We ended up with the joint ψ-function that is not factorizable and separable into two independent wavefunctions; this means that there is no quantum physical state for either particle, individually, when the ψ-functions are entangled. However, a measurement on one particle always returns a definite value, either '+' or '−' randomly, for that particle. Let us say, we get '−' for A. This immediately endows a definite quantum state of ψ_{A-} for A. The total wavefunction collapsed from $\Psi = \psi_{A+}\psi_{B-} + \psi_{A-}\psi_{B+}$ to $\psi_{A-}\psi_{B+}$. Then, we can deduce the value as well as the quantum state of the other particle B, as '+' and ψ_{B+}, *without any actual measurement*, from the only possibility left in the collapsed joint ψ-function $\psi_{A-}\psi_{B+}$. A real measurement on the second particle can then verify if the deduction has been correct, if needed.

Hold the situation in contemplation for a moment. We have two particles, separated far in space, amenable to independent measurements, possibly in different laboratories. Neither particle has any physical state or ψ-function individually, in the theory of QM. But they have a description in terms of a joint ψ-function, for the two particles together as one system. An observation on one particle A endows it with a definite state and a ψ-function of its own. Then, instantaneously, the other particle B, that has no physical state or identity of a ψ-function, spontaneously acquires a physical state! This happens irrespective of the distance between the particles. I reiterate that the whole scenario is that of the *theory* of QM, when it describes the simple physical process of two independent measurements on a pair of particles that have previously interacted. That does not imply that there are instantaneous causal influences in the actual physical phenomena; one needs to be aware of the distinction between a phenomenon and its theoretical description.

Now, as a rational physicist, suppose I insist that the physical state of a particle, whatever it is, cannot be changed instantaneously by any action of observation I do at a location that is far away from that particle. Note that I am not insisting on any requirement on

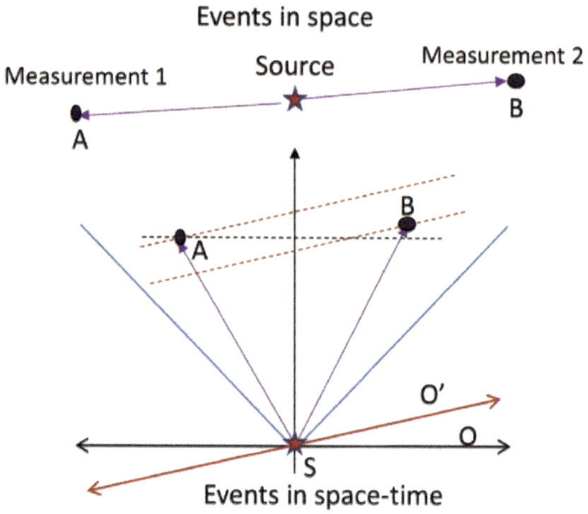

Figure 9.2 The measurements on a pair of spatially separated particles that originated in a common source. When the particles are far, one measurement cannot be the cause of the result of the measurement on the other. Any speculation of a causal relation is nonlocality, or 'spooky action at a distance'. Such a causal relation has no invariant meaning in the STR. In the frame O, event B is after the event A, but in the frame O', B is before A (the line of simultaneity of O' is parallel to the spatial axis of O')

any physical property of the particle. I am only imposing the strict locality constraint, as Einstein insisted, that there is no telepathy of physical measurements. I cannot connect the two measurements by a causal relation; a measurement and its outcome on A cannot be the cause of the instantaneous outcome of a state for B and vice versa (Fig. 9.2). But, we saw that a particle that did not have any ψ-function of its own within the theory of QM did acquire one, apparently as a result of a measurement done far away. Then, respecting the constraint of locality, one has to conclude that the QM description in terms of the ψ-function is not a faithful representation of the factual physical situation; *ψ-functions do not represent individual physical states.*

The EPR argument in 1935 was in the same spirit, and had the same content, though not as clear or transparent as the description I outlined above. In fact, Einstein lamented about the loss of clarity of the argument in the published paper, because the 'main argument was buried in erudition' in Podolsky's drafting of the EPR paper for the journal *Physical Review.* Einstein had called the inadequacy of the ψ-function of QM to represent physical states in nature as 'incompleteness'. Even stalwarts have been misguided by the wrong reading of that paper. In particular, a whole generation was misled permanently into thinking that Einstein was calling for a new classical statistical theory, in which a particle had definite values for all possible physical variables, without the characteristic superposition of physical states. These values manifest randomly during the observations because of the action of some mysterious hidden statistical variables in the environment. No doubt, Einstein would have been very pained by this attribution, which happened much after his death, because *it is easy to verify that the EPR paper does not contain any phrase that suggests hidden variables or extra statistical variables to 'complete' QM.* In any case, it is far better to read Einstein's own short version of the EPR argument, than the four-page EPR paper, to see clearly what exactly he meant to say.

The precise and complete content of Einstein's version of the EPR argument can be seen in many of his later writings. I quote from his paper 'Physics and reality' in the *Journal of the Franklin Institute*: (Einstein, 1936):

> Consider a mechanical system consisting of two partial systems A and B which interact with each other only during a limited time. Let the ψ-function before their interaction be given. Then the Schrödinger equation will furnish the ψ-function after the interaction has taken place. Let us now determine the physical state of the partial system A as completely as possible by measurements. Then quantum mechanics allows us to determine the ψ-function of the partial system B from the measurements made, and from the ψ-function of the total system. This determination, however, gives a result which depends upon which of the physical quantities (observables) of A have been measured (for instance, coordinates or momenta). Since there can be only one physical state of B after the interaction, which cannot reasonably be considered to depend on the particular measurement we perform on the system A separated from B, it may be concluded that the ψ-function is not unambiguously coordinated to the physical state. This coordination of several ψ-functions to the same physical state of

> system B shows again that the ψ-function cannot be interpreted as a (complete) description of a physical state of a single system.

That is the whole 'EPR argument'! There are three steps to the proof, (a) different ψ-functions correspond to different physical states, (b) the choice of which type of ψ-function will result for B (momentum or position, for example) is dependent on *a free choice during the measurement at A*, and (c) the physical state of a particle cannot be altered or influenced by another measurement done far away: then there cannot be complete correspondence between the physical states and ψ-functions. This is the incompleteness in QM. Nowhere in this proof, Einstein assumed that the physical state of the particle is specified by definite values for all observables. The essential argument is about the lack of one-to-one correspondence between the quantum mechanical state represented by the ψ-function and the physical state of material particles; the nature of the physical state does not enter the argument. In fact, Einstein was very clear that, in the theory of QM, neither particle had any physical state before a measurement, because there was no ψ-function for either particle. He wrote in a paper (Einstein, 1948) on 'Quantum mechanics and Reality',

> Let the entire system be completely described in the quantum mechanical sense by a ψ-function ψ_{12} of the coordinates q1, ... and q2, ... of the two part-systems (ψ_{12} cannot be represented as a product of the form $\psi_1\psi_2$ but only as a sum of such products). At time t let the two part-systems be separated from each other in space, in such a way that ψ_{12} only differs from 0 when q1, ... belong to a limited part R1 of space and q2, ... belong to a part R2 separated from R1.
>
> The ψ-functions of the single part-systems S1 and S2 are then unknown to begin with, that is, they do not exist at all.

The issue of time and simultaneity enters the scene in a very fundamental way in any scheme that attributes the status of the 'cause' to the measurement on one of the particles, and consider the second particle acquiring a physical state and a ψ-function as the 'effect', irrespective of how large their separation is. This is easy to understand after our discussion of the temporal relationships between spatially separated events for various observers in relative motion. One may refer again to Fig. 9.1. Suppose the event A is the measurement on one of the particles, which results in a definite outcome for a physical observable. Let the event D be the measurement on the other particle that is spatially far separated. The two results are correlated, and QM has the statistical description of this correlation. Both classical physics and QM predict the result of the individual measurement event D, given the result A, and vice versa, when the measurement is of the same physical quantity with identical settings of the measurement apparatus. This is by deduction using prior information on the correlation. Now, if one observer O insists that A is the cause of D because A happened before D, another observer O' who is in motion relative to O could insist with equal right that D is the cause of A, because D happened before A. For the events A and E, which are simultaneous for O, the observer would need to invoke truly instantaneous influence passing between A and E. In contrast, O' would claim that E is definitely the cause for A.

In Cosmic Relativity, the situation is not mixed up as in the STR because time and simultaneity are absolute in Cosmic Relativity. Therefore, there is no question of ascribing the cause-effect relation to such events (like A and D), by any observer.

Summarizing, there is a strong conflict between the physics of relativity (both the STR and Cosmic Relativity) and the ascribed phenomenon of nonlocal influences in quantum mechanics. But, most people believe and vouch by their conviction, surprisingly and somewhat shockingly, that there is definitely such nonlocality in the physical nature, in quantum correlations of spatially separated entangled systems. What is the reason for this unphysical and irrational belief among physicists, which reminds us of the fellow who was cutting the branch of the tree he was sitting on? This is what we shall explore briefly now.

9.4.1 The Origin of the Belief in Nonlocality

In the EPR argument, Einstein assumed strict locality, and proved that the wavefunction of QM that can be affected by a measurement done far away is not a faithful representation of the physical state of a particle. Einstein preference was to interpret the wavefunction as a representation of the statistical features of the state. This implies that there is a deeper and different theory for the description of individual quantum events and measurements, which remained elusive. In 1964, J. S. Bell *interpreted* the EPR paper in *Physical Review* as a call to replace QM with a classical statistical theory with some 'hidden variables' that are responsible for the unavoidable statistical fluctuations in the microscopic phenomena (Bell, 1964). The hope in such theories was that one could get notionally deterministic quantum physics, akin to the classical mechanics underlying classical statistical mechanics, by replacing the quantum uncertainty attributed to the wave-particle duality with mere statistical uncertainty, due to the randomness in a set of unspecified hidden variables. There is an unfounded impression, reinforced by Bell's paper, that Einstein argued for completing quantum mechanics by replacing it with a classical statistical theory with hidden variables, of the kind considered by Bell. Any such impression is a gross distortion of the content of the EPR paper and Einstein's views. The EPR paper does not contain phrases like hidden variables, statistical theory, etc., as it can be easily checked. But, Einstein was no more present to clarify that no such mission was intended (Unnikrishnan, 2021b).

In any case, Bell derived the maximum value for the correlation possible in any classical statistical theory that respects Einstein locality, with hidden stochastic variables and definite prior values for the observables of a particle (Bell, 1964). As mentioned earlier, such theories are called Local Hidden Variable Theories (LHVT). He found an upper bound for the spin correlation of two spin-1/2 particles with total spin zero. The statistical average of the correlation of the measurements on the two particles in the LHVT was bounded by a maximum value, and this relation was called the Bell's inequality. The quantum theoretical result for the same correlation in the spin singlet state (total spin zero) was known, and exceeds this upper bound. Hence, the Bell's inequality is 'violated' in the theory of quantum mechanics. *This implies that the quantum theory is certainly not a classical statistical theory*, as one knew already (Unnikrishnan, 2005a). More importantly, the experimental results on the correlation of the pairs of measurements on two spatially separated particles also exceed the Bell upper bound, violating the inequality; they in fact agree with QM, as most people expected. So, one could say that nature violates Bell's inequality, but obeys QM.

The interpretation of all these developments should have been straightforward – that the LHVT are not tenable; such classical statistical theories cannot be the replacement for quantum mechanics. But then, the unexpected happened! *Bell's theorem was the limited statement that the correlations in any classical statistical theory with strict locality and hidden variables (LHVT) were bounded by a definite upper limit.* Experimental results exceeded and violated this bound. Naturally one should have concluded that the LHVT are not valid physical theories. People who liked the LHVT for some reason could try to keep the Hidden Variable Theories (HVT) in contention, by making them nonlocal – then one could attribute a hidden communication and conspiracy between the particles to increase the correlation beyond Bell's upper bound, in principle. That is, a version of the hidden variable theories with some mysterious and instantaneous communication between the particles could be speculated as a new theory to replace QM. Adding such telepathic 'nonlocality' to the hidden variable theories is just a theoretical exercise in the HVT, with no implication to either the theory of quantum mechanics or to what is happening in the empirical nature. However, for inexplicable reasons, many, including J.S. Bell, started claiming that the quantum phenomena in nature violate locality! The belief that there is indeed telepathic communication between particles became widespread, despite the fact that the experiments showed only that the HVT are not tenable as long as they are local! There is no logical path whatsoever to reach the conclusion of nonlocality in QM or in nature, starting with the EPR argument, Bell's theorem, experimental results, etc. Yet, the terrible misinterpretation happened, and stayed.

9.4.2 Are There Nonlocality and Telepathic Influences?

NO! no! nO! No! It is important to deny the violation of locality in physics in as many ways as possible because there is utter confusion and serious misunderstandings on this matter. Let us clear this confusion once and for all. There is no evidence, either logical evidence in the theory or empirical evidence from any experiment, that there is any nonlocal influence in quantum phenomena. The only relevant result is that the correlations of measurements on two (or more) particles cannot be explained by a general classical statistical theory with hidden variables, under the assumption of Einstein locality. It is this judgment on a class of flimsy speculative theories with an already unphysical bag of unobservable hidden variables that was hailed as the experimental proof of nonlocality in nature!

It is quite a disappointing matter that most physicists, of all people, today believe that there is indeed some kind of undetectable and mysterious telepathic communication happening between spatially separated physical systems of the microscopic world. Indeed, the situation has become so bad that the term 'nonlocality' (which is the violation of Einstein locality) is used as a synonym for the technical term 'entanglement' in quantum mechanics!

It is worthwhile to quote two important results, shown relatively recently, in this context. One is that it is a simple matter to prove that the local hidden variable theories do not respect the fundamental conservation laws of physics, like the conservation of angular momentum (Unnikrishnan, 2005a). Hence, they are unphysical theories that should not have been taken seriously in the first place. The second result is the explicit demonstration that the correct quantum correlations emerge from the wave-particle association and prior correlation, respecting strict locality (Unnikrishnan, 2018b). The observed correlation is the result of the quantum interference that respects the strict time order of events and relativistic locality. There is no place in physics for the superstition of nonlocality.

9.5 Summary

We have examined the concepts of locality and causality in physics and their relation to the physical time. We discussed how and why the assignment of any form of causal relation of two events that are arbitrarily separated in space is in direct conflict with the fundamental tenets of relativity and simultaneity, be it the STR with relative time or Cosmic Relativity with absolute time.

I discussed the conventional role of time in the theory of quantum mechanics. Then, the formal aspects of the representation of physical states and the peculiarities of measurements (observations) on quantum physical systems were explained as a prelude to the complete and accurate description of the Einstein–Podolsky–Rosen (EPR) theorem on the incompleteness of quantum mechanics, under the assumption of locality. Einstein's own clarificatory writings on the topic were quoted and discussed. I explained in detail how the mistaken impression that 'nature violates locality' has crept into mainstream quantum physics. The door to this impression was J. S Bell's result that there is an upper limit to the correlation between the observables on a pair of particles in classical statistical theories with a set of hidden stochastic variables (LHVT), under the assumption of locality. These theories have no relation whatsoever with the theory of quantum mechanics. When experiments showed that the measured correlations exceeded the upper limit of the LHVT, violating the Bell's inequality, instead of drawing the straightforward conclusion that the LHVT are not tenable theories of natural phenomena, the strange and unwarranted conclusion that ascribed nonlocality to physical nature and quantum mechanics was advanced. The conclusion that a measurement at one location can causally influence instantly the result at another location that is arbitrarily far away was preferred and advocated by many, for reasons beyond logic and facts. This superstition of telepathic nonlocality has grown enormously. But, a systematic examination of the facts reveals easily that there is not even the slightest experimental indication of nonlocality in any part of physics. While it is difficult to say whether the superstition of nonlocality would ever wither out, the matter-of-fact road map on how the unfortunate situation was reached seems enough to stay clear.

Earlier I quoted Einstein's sarcastic mention of a conversation with an important physicist regarding the widespread attitude towards locality in quantum phenomena,

> He: "I am inclined to believe in telepathy." I: "This has probably more to do with physics than with psychology." He: "Yes."

It is unlikely that Einstein made up the situation. In any case, this is what one can expect in a typical conversation today, though the word 'telepathy' would be replaced with 'nonlocality'. The implication is that there is some instantaneous communication between the particles, which remains inaccessible to us. In a major conference dealing with the foundational problems of the quantum theory (Bari, Italy, 1985), Karl Popper had declared, 'the end of locality would mean the end of rationality in physics'. The totality of facts discussed in this chapter should help to defend the rational pillars of locality and causality in physics.

Chapter 10

Apparent Arrows of the Corporeal Time

In the previous chapter, we dealt with causality and the order of physical events in time, in the specific context of quantum mechanics and quantum measurements. The sequence of events with chains of interlinked causal order suggests a flow of time with an apparent predetermined direction. It is the general notion and impression of the directed time that we explore in this chapter. The insights on the links between time, gravity, and the cosmos will help us to understand better many widely debated issues.

The hypothetical unidirectional flow of time cannot be defended well within modern physics, surprisingly. How can one even hope to defend it after postulating the unified picture of a space-time, with equal and similar status for both space and time?! There is no flow of space, nor is any 'arrow of space'. Therefore, *any statement of a continuously flowing physical time can only be within a physical worldview that recognizes a clear separation between space and time*. We will see that the impression of directed time is primarily psychological, rather than physical.

Another issue that is often discussed, and of immense interest to the whole humanity, is the 'possibility' of time reversal and time travel. These speculations are closely linked to the psychological

Gravity's Time
C. S. Unnikrishnan

ISBN 978-981-4877-72-5 (Hardcover), 978-1-003-25656-4 (eBook)
www.jennystanford.com

impression of the flowing time with real extensions of the present to the past and the future.

The notion of self-propelled time, moving independent of matter, space, and life, is a universal image in the human psyche. We recall Newton's description in Principia, '. . .time, of itself, and from its own nature flows equably without regard to anything external. . .' (Chapter 1). The psychological image of flowing time, from the past to the future while we witness the present, is so strong and natural that 'time' is decorated with numerous equivalent imageries. An 'Arrow of Time' is a very suggestive and apt image for the experienced time. The image represents qualities like 'eternal motion', 'unstoppable' and 'irreversible'. The other implicit properties like a 'uniform' and 'fast' passage are secondary. In any case, physicists think that the unidirectional arrow exists in practice, and it needs a physical explanation. It is a hard and deep problem because all physical processes in their elementary atomic form, as represented by the basic equation governing physics, are invariant under time reversal (there are subtle and rare exceptions involving unstable elementary particles, but they are not relevant in the general discussion). This means that they remain as they are, with exactly the same appearance, if the time variable 't' in those equations is changed to '$-t$'. If the equations are invariant under time reversal, then the physical consequence they represent also would be time-symmetric. All searches for a gross time asymmetry in the fundamental interactions that govern most physical processes, especially those that matter in the human life and experiences, ended in vain. Then, why do we see a manifestly time-asymmetric world in which we cannot even take back an uttered word that disperses through the surrounding air? Why is it that we do not have any real experiences of the spontaneous reconstruction of a shattered glass or the natural revoking of a spreading stain of ink?

This question was answered satisfactorily only in the early 20th century. Before we look at that answer, we should examine the notion of the 'arrow of time' and its several avatars. We will always keep in mind that 'time' is just another name for the physical processes in localized matter, which are connected across spatial locations through physical interactions (Chapter 2). This firm and well-founded wisdom will help us to answer the most intricate questions about the postulated flow of time.

10.1 The Arrows of Time

By an 'arrow of time' we mean a directed passage of time, with the direction remaining steady and generally irreversible, as if it was determined by some fundamental physical considerations. There are several arrows of time that have been discussed in science as well as in philosophy (Reichenbach, 1971; Zeh, 1989). Most prominent are the 'thermodynamic arrow of time', the 'cosmological arrow of time', the 'quantum mechanical arrow of time', and the 'biological (evolutionary) arrow of time'. One may also discuss a 'psychological arrow of time', which is linked to the cumulative nature of our memory. Since our topic of discussion is the relation between gravity, cosmology, and time, and not the general nature of time, we will discuss only certain points relevant to the main theme.

It is easily noted that these arrows refer to lots of matter and to the collective and average behaviour of matter, be it a collection of molecules of gas and liquids, the stars and galaxies in the Universe, the large number of molecules and their aggregates in living systems, or the innumerable neurons in our brain. So, these arrows do not directly address the elemental time, which is supposed to be existing independent of matter. The arrows of time refer to the manifestation of time through the collections of elementary forms of matter.

The most important feature that is required for the mutual consistency of these various arrows of time is that they should be directed similarly, from a common past to a common future. In other words, the conceptual arrows and their manifestations should be isomorphic. The physicists should be able to prove that this indeed is the case, lest there be any lingering doubt or fear that the unexpected behaviour of time arising from the clash of differently directed arrows of time might disrupt our regular existence. The coherence of the different arrows is important in attributing the property of 'flow' to time.

In addition, we would like to know the relation of these material arrows to the pure elemental time, conceived independent of matter and space in conventional physics. In this task, we will fail if we do our search within conventional physics, because of the reason already discussed; such an elemental time does not exist. But, owing to the sound insight on the nature of the physical time, as synonymous with the localized evolution of the states of matter, our answers will be reliable.

10.1.1 The Thermodynamic Arrow of Time

This is a technical name for the familiar manifestations of time in our experience; growth, ageing, spilled coffee, uttered words, spreading and dissipation of waves in a pond, pandemics, climate changes. . . , etc. In all such phenomena, there are two characteristic traits: one, the physical system consists of large number of elementary material constituents, and two, the system is progressing from a state where it shares its resources in a smaller number of underlying possibilities to a state with a much larger number of shared possibilities. *The number of possibilities in which a resource like energy is shared with equal probability is the abstract thermodynamic quantity called entropy.* This is a conceptual quantity and not something that can be sensed and measured with an instrument, like temperature or pressure. Thus, *entropy keeps increasing in an isolated system, till a maximum, and this statement is known as the 'second law of thermodynamics'.* In physical terms, entropy is the measure of the number of shares of an internal resource like the energy among the material subsystems, like the atoms in a gas or the cells in an organism. Since randomness and fluctuations have a role in such sharing with equal chance, one often equates 'disorder' with entropy. Only time symmetric physical laws are at play, at every stage of the underlying physical processes, but the asymmetry in the number of possibilities in sharing the physical resources, ever increasing, propels the arrow of the physical evolution. The impression of the arrow is an averaged notion, ever fluctuating in its course, because of the large number of particles and their physical states involved. The fluctuations become relatively insignificant as the number of elementary material entities and the possibilities of their physical states become large.

Consider just a few particles in a small room, free to move about and collide with each other. If we start with an initial configuration in which the particles are together in a corner, soon they will spread all over the volume of the room. But, there is also some chance that once in a while they would come together in a small region near the same corner they started from. Similarly, if one could reverse the motions of all the (very few) particles at any stage, the general dynamics would look similar to the original scenario. There is nothing at all to ascribe any asymmetry in the physical evolution in this case.

However, if we consider millions (denoted as 'n') of particles, all starting together in a corner, the experience would be very different. *An arrow of time emerges*, even though the physical laws remain exactly the same. The configuration that started in a corner evolves quickly to a more or less uniform appearance, with particles everywhere, which remains without noticeable changes. The particles move about and share their individual energy and momentum in the numerous collisions. Now, the only difference from the previous one is in the number of particles and the consequent exponential increase in the number of possibilities in the positions and the velocities of the particles. Let us denote it by N. All possibilities have an equal chance in the room. But, only a very small number of these possibilities correspond to the state, 'being concentrated near a corner' N_C. Most other combinations of the positions and the velocities correspond to the particles nearly uniformly spread in the volume of the room. The simple fact that $N_C \ll N$ makes the possibility of getting together in a corner impossibly small, but not zero. This is the origin of the experience that an ordered and simple physical state (of starting together in a corner) naturally and 'irreversibly' evolved into a state of total disorder that filled the available space. This apparent irreversibility happened without any intrinsic time-asymmetry in the physical laws.

As a 'visible' example, a drop of ink in a glass of water is a special 'initial' physical state that we can prepare. But due to the molecular collisions and rearrangements in the liquid, the molecules of the ink are shared to more and more spatial locations. Each location's neighbourhood is like the original drop, only thinner. The random sharing goes on till the ink is uniformly spread. The number of shares keeps increasing, with the value of each share correspondingly lower. Entropy is increasing. The chance that the numerous random collisions result in the clumping of some ink to a denser droplet is impossibly small, but it is not zero. *To destroy order, the disruptive agents can act independently and randomly, but to restore order everything needs to be coordinated to act towards a single goal.* The asymmetry inherent here is the source of the thermodynamic asymmetry. When everything is shared more or less equally, no further increase in sharing is possible, though sharing back and forth is going on. Entropy reaches a maximum value. *Thus, the thermodynamic arrow is the result of starting in a special low entropy state. The asymmetry is between the initial and the final states, and not in the elemental time.* It is important to realize that the physical processes in this case (like the spreading ink) themselves qualify as 'time' and a legitimate 'clock', though not very convenient to use due to the non-uniform evolution.

Another common example is that of a wave, of sound, light, etc., propagating out from a source, transferring a portion of its energy to various absorbers on its way. A portion of the absorbed energy turns into internal heat in the absorber, its molecular motions, and the rest is emitted further as secondary waves. The time-reversed version can happen only if all these myriad material absorbers emit back the right amount of waves, even converting their internal heat back to the waves, to build a coherent wave that goes back to the source. This requires the coordinated and coherent action from the innumerable number of the uncorrelated molecules at different locations. But physical processes are not known to have a purpose. It is easy to see that the probability for this to happen is vanishingly small, deserving only the adjective 'impossible'. Thus, the origin of the arrow of time in these cases is clearly traceable. They are really the arrows of interacting physical processes.

A seed sprouting to become the trunk and then the many branches of a tree is an arrow of time that is launched from a relatively simple initial physical state. The tree with its many branches results from the same kind of repeated replication of

Figure 10.1 The growth of a tree represents time itself, and also a directed arrow of time. All the elementary physical processes in its cells are reversible (time symmetric), but the becoming of the tree is practically irreversible. The cause is beyond the roots, of starting with the relatively simple (low entropy) initial state of a seed. The ordered growth with low entropy is maintained, despite the large increase in the number of participating cells, with constant input of energy.

molecular cells, each involving an element of chance (probability). The random sharing and the increase of the thermodynamic entropy are avoided in the ordered growth only with a constant input of energy. In other words, a living tree is not an isolated physical system, subject to only the second law of thermodynamics (Fig. 10.1). Even in such complex systems, the elementary physical processes and many chemical processes are reversible. Yet the mature state of the tree is irreversible.

Thus, there is no arrow or directionality of elemental time in the thermodynamic arrow of time. It is the situation of a special initial state, with a relatively small number of shared possibilities, that develops through time symmetric processes into a vast number of shared possibilities; the arrow slows down and stops eventually, when the possibilities to share are saturated.

10.1.2 The Cosmological Arrow of Time

This is the arrow of time suggested by the ever-expanding Universe. The fact that the galaxies are receding from each other is called the expansion of the Universe. At present, the current theory of cosmology, based on the General Theory of Relativity, does not favour or predict expansion over contraction; thus, the expansion is merely an observational fact, taken as 'given'. However, the observed expansion provides a 'clock' and a notional direction of time.

One may ask whether time would reverse if the Universe started contracting! That would be weird. After all, a stone that is thrown up fast keeps going, with progressively decreasing speed, until it stops and reverses its course while falling back. But, nothing happens to the progress of life and ageing in this world, even if thousands of thrown stones reverse their motion. The Universe is a similar dynamical system, in the theoretical description. There is no relation between the expansion of the Universe and the perceived unidirectionality of time. What is important for the impression of an arrow of time is that the contents of the Universe undergo changes that are irreversible. For example, clouds of matter collapse to form stars, they burn nuclear fuel to generate radiation that goes light years away powering other phenomena and even life, stars collapsing to their eventual death to black holes, etc. It is the totality of these phenomena that are practically and naturally irreversible, which makes the cosmological arrow. Thus, the cosmological arrow of time is really the expansion of the Universe superposed on a mixture of many thermodynamic arrows of the thermal and gravitational processes in the Universe. The fact that the Universe keeps expanding, while all such phenomena go on, is just an indicator. A contracting Universe will not make a shining star out of a black hole. The nature of gravity prohibits that.

It is in the cosmological arrow of time where gravity plays a decisive role. This is because all large-scale phenomena, like the formation and evolution of the stars as well as the collective dynamics of the whole Universe, are controlled by gravity. Besides, gravity provides a unique and truly irreversible physical phenomenon – that of the formation of black holes. A massive star that has exhausted all its nuclear burning will collapse. The time-reversed process is the reversed dynamics of an expanding star that starts burning nuclei in the reverse order. However, this becomes a true impossibility if the star has already collapsed to the stage where it is a black hole, because reversing dynamics of matter in a black hole requires exceeding the speed of light. Thus, gravity plays a special and decisive role in the fundamental irreversibility of time.

10.1.3 The Quantum Mechanical Arrow of Time

There is a distinct difference between the past and the present in quantum mechanics. This is related to the foundational traits of quantum mechanics, manifested in the uncertainty principle and the quantum measurement. The Schrödinger equation for quantum dynamics is time symmetric, in the sense of the equation being a differential equation for a smooth evolution (time reversal in quantum mechanics is technically more involved than just reversing the sign of time from $+t$ to $-t$ in the equation, but we do not need to discuss that detail here, in our brief overview). But the act of observation of any physical quantity interrupts this smooth evolution and effaces the memory of the evolution till then. This is the troublesome 'collapse of the wavefunction' that we encountered in the previous chapter. A new evolution starts, without the memory of the past physical state, to which one can never return. It is important to note that the quantum mechanical arrow is not related to any complex behaviour of a multi-particle system. The irreversibility associated with a measurement applies to the dynamics of a single particle.

A more transparent way of analysing the quantum mechanical arrow of time might be in terms of its characteristic feature of the probabilistic stochasticity in the elementary processes. The effective irreversibility in quantum phenomena can be seen even in the case of a single radioactive decay or in the emission of a photon from an atom. The emission of a radioactive product is a random process involving chance, and tunnelling through an energy barrier. The time-reversed version will also need to face this barrier again and the associated chance element. Hence, time-reversed dynamics does not result in getting back to the source because of the much smaller double-chance involved. Similar factors can be seen in other quantum mechanical processes as well.

10.1.4 Arrows of Time in Life

The psychological arrow of time is the one that has provoked most of the questions we ask about time. The psychological arrow is closely and inseparably linked to perception, thoughts and 'memory' and hence, material locations (in the brain) that can hold the elements of our perception and thoughts. We will call the collection of these locations and their contents together as memory (for a simpler discussion). All stored records of past events classify as memory. I emphasize that the retrieval and the manipulation of the contents of the memory require the special 'shelves of memory', or 'registers' where the contents of the memory can be shifted and moved in and out at will (where the 'will' itself would need such special registers). These registers are finite in number. Therefore, a filled and unchangeable stack of memory registers implies inaccessible records, and marks the end of perception, thoughts, and history. However, such a physical limitation is not perceptible because the number of the memory locations in the brain is very large, and they seem 're-writable'.

The psychological arrow is different in its character and in its operational structure from the thermodynamic arrow, because the filling of the possibilities of physical states of memory happens mostly by a 'conscious' effort, rather than by the uncontrolled and unsupervised action of the physical interactions within the system. Similar to the other arrows, the psychological arrow also starts in a special initial state, with its resources more or less empty in memory and history, except those that are genetically determined. Instead of going from the special ordered initial state of low entropy to a mostly unordered state of high entropy, the psychological arrow of time progresses through the ordered expansion of the memory states, resisting and guarding against the increase of disorder entropy. *This requires a constant input of energy*, to keep the disorder and randomness low at all times. Therefore, the psychological arrow of time has its basis in the 'open' and non-isolated nature of the physical system that allows such input, as required. The similarity with the living tree may be noticed, in spite of the obvious differences in the structure and function of the systems.

All the natural questions we commonly ask about time are dressed by our experience with our individual as well as collective psychological arrows of time, which definitely have a 'past' and a 'future', apart from the 'now' we are sure of. However, we should realize that this past has a localized existence only as our memory. Likewise, the future resides in similar locations of our memory, as projections of our mind. Both the past and the future reside *here*, in our memory, *now*. Whichever way we do our analysis of our psychological time, this realization is inevitable. I do not know whether one would find such realization about time as an anticlimax or as a relief. But, it is surely a valuable wisdom.

The psychological time based on our memory registers in the brain is the only kind of time along which one can freely do time travel, either consciously and voluntarily or otherwise, without bothering about the physical constraints of continuity or causality. That cannot be confused with any form of physical time travel. The example is instructive because it also sensitizes us to the localized material nature of time, with all the past existing in the present and here (in our brain, in this case).

10.2 Where Is the Bow?

Though we can identify clearly the passage of the arrow of time in the numerous examples involving a large number of elementary physical systems, the fundamental 'bow' that propelled the arrow of time remains invisible. I mentioned that the thermodynamic arrow is actually the result of starting in a special low entropy state and is not related to any intrinsic property of time. Still, we can explore a bit further the interesting question whether the elemental time has a direction, as it was universally thought before the present understanding of the thermodynamic arrow of time matured. If the elemental time is indeed unidirectional and irreversible, what could be the reason?

A justification can in fact be found in the depths of fundamental physics, even though it is an abstract connection. To understand it, we may go to the simple example of the motion of a particle in space. What drives its motion? The initial impetus loads a certain amount of momentum – the product of mass and velocity – in a body and thereafter the body moves perpetually. One says that *the momentum is the generator of the spatial translations.* Similarly, the angular momentum is the generator of rotations. This is a literal connection, but this is formally expressed in the mathematical framework of mechanics and the same term 'generator' is formally used. The question 'what is the generator of translation in time?' is then formally relevant. There is a definite answer in physics; *the generator of time translation is energy.* So, if there is any bow that propelled the arrow of elemental time, it should be related to energy in a fundamental way.

Now we have a very interesting situation. We can move in both directions in all three dimensions in space because its generator, the linear momentum, can have both signs, positive and negative. However, energy is always positive (though changes in energy can be negative), if all contributions are included. Therefore, it is reasonable to assert that *the elemental time is unidirectional because the generator of translation in time has only one sign.* The interesting consequence in formal physics is that one cannot reverse the translation in time unless energy becomes negative.

There are extensive theoretical investigations regarding the absolute positivity of energy. Those discussions are too formal and technical for a mention here. However, we note that energy, like mass, is the source of gravity. Therefore, the intimate relation between energy and time is also the intimate relation between gravity and time. Based on these connections, theoretical physicists have advanced various speculations on time and its reversal (Morris and Thorne, 1988). However, all such proposals are set in an imaginary world that is far removed from the real physical world with the vast cosmos and its matter in the background, perennially under the gravitational influence of the cosmic matter.

10.3 The True Nature and Essence of Time

It is perhaps important to review the real nature of time to make sure that we understand it completely. Then we will be able to answer with conviction many questions that have been bothering physicists and philosophers for ages. The first step in gaining this understanding is to shed all prejudiced intuition about time, ingrained due to our extreme familiarity with matter and space.

Observe the manifest nature of time – there is only the present, at this moment. We have been counting simultaneities with a reference clock and what we have at present is the number of 'nows' that are accumulated plus the NOW of present. These numbers are updated, and the memories are updated, and we tend to consider it as the past that exists physically in time. We justify our inability to be physically in the past by attributing the property of irreversibility to time! But, this inability arises really from the fact that there is no existence for any physical entity called the past. There is no physical past other than the updated number of simultaneities with local clocks (other physical processes), or 'time-numbers'. Neither is there any future existing now; that is the creation of the free projection of our imagination. All that exists is a number that quantifies the simultaneities, stored in a physical memory that is present in the local space NOW, at our location. Each individual or a piece of matter has only this number counts and the local NOW. These numbers at different locations will have a relation only if there were interactions between these different pieces of matter. The relative ordering and causal relations of the time-numbers at different 'NOWs' happen only through interactions. This is what we call time.

In contrast, we imagine the 'past' as something that exists because of two reasons. One is the *memory of the past that exists* in our brain, though without our conscious awareness that it is a mere record that exists 'inside' us. The other reason is our comparing time with what we experience of the space and matter, which have the property of extension. I, at the location Mumbai, am aware that my friend is in Paris and my mother is in Kerala. I can verify this by moving from my location to the other locations. However, what I did yesterday to now is only a record, a number, that exists now at my location. There is only the time of NOW. Yesterday has no physical existence. Therefore, I cannot move to yesterday, or tomorrow. Time is not a flowing river, in which one has to only swim back hard enough to reach the still flowing past!

While drawing pictures of space and time together in physics discussions, we tend to draw a continuous 'axis' for time, extending from the past to some future, replicating what we can verify about space that is extended. Not for one moment we stop to think and realize that there is no such axis. For each location, there is just a number representing counted simultaneities, appropriately recorded in a physical memory (paper, electronic, neural, etc.) – that is the only reality of time, whether physical or psychological. The axis of time is factually only an infinitesimal point at each location. It will distract us from the main deep insight if we start debating whether this point is a mathematical point or whether it has some tiny extension. We can choose either, to suit our logical mind, but *it is important to focus on our discovery that the only entity that exists as time is the accumulated counts of simultaneities at our location, NOW.*

Isn't it obviously naive and unphysical then to think that one can reverse time, or do time travel, etc.? One cannot reverse the number of simultaneities. One can only accumulate more simultaneities and store the new number as the time of the new NOW.

We can connect the different time-numbers at different locations by sending signals (interactions) and synchronize them, if we please. However, it is very important to realize that our time is local and consists of only the NOW and a number, representing the past simultaneities with our local reference clocks (Fig. 10.2).

10.4 Is Time Reversible?

We can now address some deep questions, the answers of which we already know, at least provisionally. The question whether time can be reversed and whether time travel into the past or future is possible, passes anybody's mind, as a universal and fundamental question. This happens because of our intimate awareness about the passage of time and our associating the daily pleasures as well as our difficulties with the passing time. However, these questions are usually treated without adequate physical analysis, or within the narrow premises of expert disciplines. Typically, the physicist is seen as the ultimate technician and judge to evaluate and implement, if possible at all, time travel and time machines. After realizing that the past does not have any physical existence, a moment's thought will tell us plainly that the speculation does not make physical

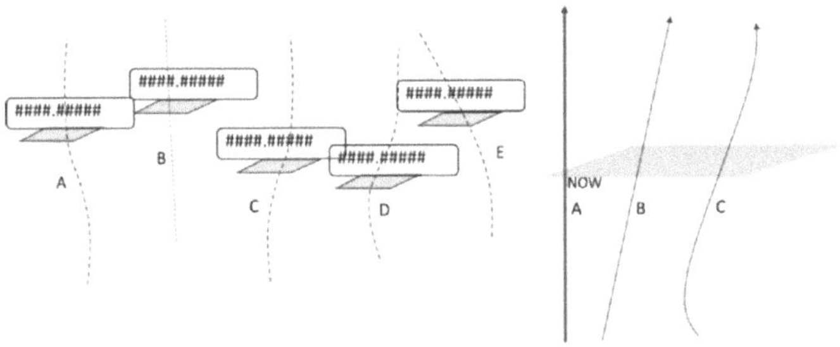

Figure 10.2 The left panel depicts the real factual situation about 'time', revealing its true nature and physical content. There is only the present, locally, and the physical existence of the entire past is merely as the numerical record of the elementary simultaneities of the past physical processes with a reference clock. Each record, A, B, C, etc., is *independent and local*. When the records are connected by interactions, a global notion of time emerges. On the right is the conventional imagination about the flowing time, as something that really exists physically, from the past to the future, akin to the extended space. Present (now) is treated as one point in the continuous stream of time.

sense except in the simplest dynamical sense of reversing the simple motions of a few particles. Most proposals of physical time reversal are associated with mere motion reversal.

But, take the case of time reversal for a human being. Time reversal at any moment would mean reversal of not just the external time and physical events, but also everything associated with life, from mundane bodily functions to complex cellular functions. To sensitize to the problem, the simple fact of reversed blood flow and metabolism might do – life in the present form will become impossible, killing the adventure project without any progress. (In fact, time-reversed life could be quite disagreeable!) In usual discussions, these aspects are not remembered because physicists think in terms of the dynamics of lifeless single particles, for which time reversal can be identified with the simple reversal of dynamics.

There is a larger problem with the usual speculations of the possibility of time reversal and time travel. We live in a dynamical and expanding Universe, with matter and its physical process. The mathematical proposals of time reversal have to set up a Universe with very special conditions; this could be some kind of exotic matter that is not found in this Universe, global rotation of the Universe along with negative energy, etc., or special topologies of space and time. Even if one grants all that, that is not our Universe and what can be affected is not our time! We have access to only this Universe. All our life and scholarly explorations are done in this 'once given' Universe (as E. Mach put it). Therefore, the exotic speculations that one finds on papers are all misplaced.

It may be important to state the obvious about the inconsistency of any proposal of time reversal in the matter-filled cosmic environment. If one speculates on a global time reversal, then the entire Universe and all its physical processes also have to reverse in time. Already this is impossible due to the existence of collapsed stars as black holes. The matter that has collapsed to a black hole is truly irrecoverable, and hence, the phenomenon is irreversible even if all the rest of dynamics are reversed, as long as gravity remains attractive. Therefore, including the Universe and gravity in the analysis, as one should, makes all time reversal projects implausible.

In any case, the strongest argument that makes any proposal for time reversal or time travel invalid is the physical nature of time, as mere evolving states of matter. Then the accumulated time is identical to the accumulated quantity of the progressions of states, which keeps monotonically increasing irrespective of the sense of the detailed dynamics. Thus, *both time reversal and time travel are the products of our psychological prejudices about the nature of time.*

10.5 Antiparticles as Time-Reversed Particles?

Antiparticles are charge-images of particles; identical copies except that their (electric) charge is of the opposite sign. They have the same mass, and bringing a particle and antiparticle together results in their annihilation to two photons, each with the energy equivalent to the mass of each particle. In the initial days of their theoretical discovery, P. Dirac had imagined them to be equivalent to the 'absence' of a particle in a dense and infinite sea of particles. This was of course a crude and inconsistent picture because of the implied infinite mass of the sea. If one asks for any evidence for such

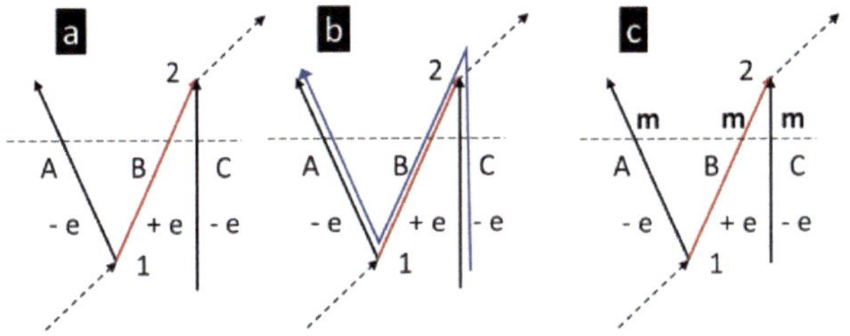

Figure 10.3 Gravity pulls down Feynman's speculation of anti-particle as a particle travelling backward in time. (a) Pair creation at point 1, and annihilation of the positron with another electron at point 2. Three particles are present simultaneously, with the total charge –e. (b) Feynman's reinterpretation as a single charge –e arriving along C to 2, and then travelling backward in time from 2 to 1 along B and then progressing forward in time along the worldline A. (c) The 'gravity gap' in Feynman's proposal. Though the three worldlines A, B, C can be considered consistently as one in which a single electric charge travels forward, backward, and then forward in time, the total mass is always $3m$, and not m. Then the gravity of the particles corresponds to the source mass $3m$, and not m of a single charge!

a Dirac sea, one has no choice but to go out of the boundaries of science and admit glibly that it is an invisible sea.

Twenty years later, Richard Feynman imagined the antiparticles as particles travelling backwards in time. His paper on the theory of positrons in the journal *Physical Review* (Feynman, 1949) has the following description (see Fig. 10.3),

> In the approximation of classical relativistic theory, the creation of an electron pair (electron A, positron B) might be represented by the start of two world lines from the point of creation, 1. The world lines of the positron will then continue until it annihilates another electron, C, at a world point 2. Between the times t1 and t2 there are then three world lines, before and after only one. However, the world lines of C, B, and A together form one continuous line albeit the 'positron part' B of this continuous line is directed backward in time. Following the charge rather than the particles corresponds to considering this continuous world line as a whole rather than breaking it up into its pieces. It is as though a bombardier flying low over a road suddenly sees three roads and it is only when two

> of them come together and disappear again that he realizes that he has simply passed over a long switchback in a single road.
>
> This over-all space-time point of view leads to considerable simplification in many problems.

Feynman mentioned the general logic, apart from the convenience and simplicity of the diagrammatic interpretation, in the same paper. The attitude is that of an explorer examining the space-time record of completed events, with freedom to treat time as equivalent to space; *the temporal order is treated as irrelevant.*

> Here we imagine the entire space-time history laid out, and that we just become aware of increasing portions of it successively. In a scattering problem this over-all view of the complete scattering process is similar to the S-matrix viewpoint of Heisenberg. The temporal order of events during the scattering, which is analyzed in such detail by the Hamiltonian differential equation, is irrelevant.

These passages contain the fundamental basis of his theory of relativistic quantum mechanics and the very useful and popular calculational tool of the Feynman diagrams. However, the conceptual edifice is fragile and breaks down in the face of the physical reality, which includes gravity and its source, the mass. Gravity alone shows that treating antiparticles as particles travelling backward in time (following the charge rather than the particle) is not consistent. Referring to Fig. 10.3, it is clear that though the total charge at any instant corresponds to one particle of charge $-e$, the mass is $3m$, corresponding to three particles. Therefore, the situation cannot be reinterpreted as a single charge traversing a zigzag continuous path in space-time. The Stückelberg-Feynman view of antiparticles as particles travelling backward in time cannot be maintained in a world with gravity. Surprisingly, the myth that the view is consistent with the rest of physics has survived without critical scrutiny till now. Particles or antiparticles, their existence is only in their local present. In fact, in our Universe with its universal time and absolute simultaneity, the temporal order has an invariant meaning. However, this obviously does not diminish the computational value and convenience of the diagrammatic techniques introduced by Feynman.

10.6 Closed Loops of Time

The geometrical representation of the physical world, and its success as a universally accessible image in the physical theories, are both the boon and the curse for our creative imagination about the physical notions. Once we see 'time' as a geometrical line that keeps on extending, as a flowing 'coordinate', the imaginative child in us would want to make his best graffiti from it. How about a simple loop? That is easy to draw and then it mathematically represents a hypothetical history of evolution (Fig. 10.5). While moving forward in time, we have gone to the past and then retraced our life, living the closed repetitive time.

The cyclical time has been considered in the intellectual and mythical traditions of several civilizations. In most intellectual and philosophical traditions, the cyclical time merely means recurring histories. However, in physics, it is related to a mathematical possibility arising from the representation of time as a geometrical dimension similar to that of space. If such space-time has peculiar geometric features, then one can end up with the progression in time that goes in a direction different, and even opposite, to the 'normal' direction, without reversing and tracing back the elapsed time. In other words, one could imagine a mathematical loop of time, that joins back the regular time trace, making a new kind of cyclical time, yet always going forward. Rather strange, but mathematically representable and consistent.

The discussion of closed loops of time was revived after the mathematician Kurt Gödel showed that one could construct solutions of Einstein's equations of the GTR that admitted such 'time like' loops. I cannot explain this more lucidly than Einstein himself, as done in his commentary notes in the Schilpp volume *Albert Einstein: Philosopher-Scientist* (Einstein, 1949),

> Kurt Gödel's essay constitutes, in my opinion, an important contribution to the general theory of relativity, especially to the analysis of the concept of time. The problem here involved disturbed me already at the time of the building up of the general theory of relativity, without my having succeeded in clarifying it. Entirely aside from the relation of the theory of relativity

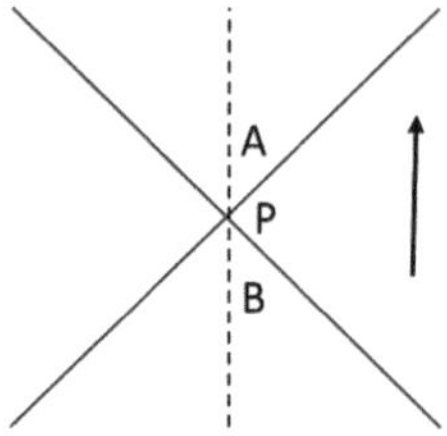

Figure 10.4 Is the time-ordering of events separated by a short duration (in proximity) preserved when the events are arbitrarily apart in duration? Figure redrawn from Einstein's commentary to the articles in the reference (Einstein, 1949).

to idealistic philosophy or to any philosophical formulation of questions, the problem presents itself as follows:

If P is a world-point, a 'light-cone' ($ds^2 = 0$) belongs to it. We draw a 'time-like' world-line through P and on this line observe the close world-points B and A separated by P. Does it make any sense to provide the world-line with an arrow, and to assert that *B is before P, A after P*? Is what remains of temporal connection between world-points in the theory of relativity an asymmetrical relation, or would one be just as much justified, from the physical point of view, to indicate the arrow in the opposite direction and to assert that *A is before P, B after P*?

In the first instance the alternative is decided in the negative, if we are justified in saying: If it is possible to send (to telegraph) a signal (also passing by in the close proximity of P) from B to A, but not from A to B, then the one-sided (asymmetrical) character of time is secured, i.e., there exists no free choice for the direction of the arrow. What is essential in this is the fact that the sending of a signal is, in the sense of thermodynamics, an irreversible process, a process which is connected with the growth of entropy (whereas according to our present knowledge, all elementary processes are reversible).

If, therefore, B and A are two, sufficiently neighboring, world-points, which can be connected by a time-like line, then the assertion: 'B is before A' makes physical sense. But does this assertion still make sense, if the points, which are connectable by the time-like line, are arbitrarily far separated from each other? Certainly not, if there exist point-series connectable by time-like lines in such a way that each point precedes temporally the

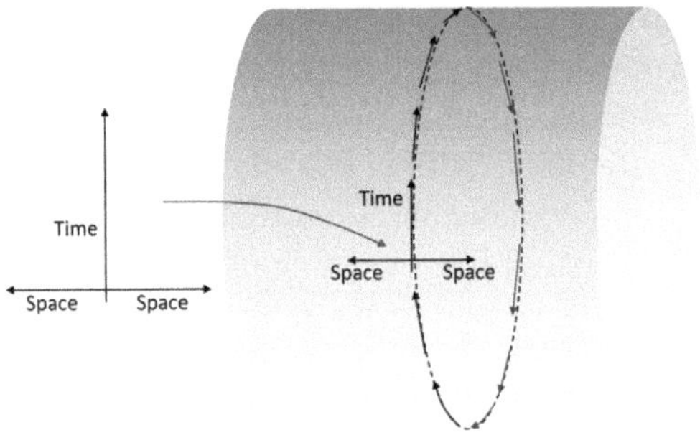

Figure 10.5 Locally 'normal' space-time (left) can have a globally cyclic time, if the space-time has peculiar global features (the topology of a cylinder that could be infinite in the spatial direction, for example). There is a closed loop in the time direction, but there is no time reversal. The unphysical aspect of a 'dimension of time' is clear here as well, because the representation of the 'future' and the 'past' as physically coexisting with the present is conceptually and logically a flaw, and certainly empirically unsupported.

> preceding one, *and if the series is closed in itself.* In that case the distinction 'earlier – later' is abandoned for world-points which lie far apart in a cosmological sense, and those paradoxes, regarding the direction of the causal connection, arise, of which Mr. Gödel has spoken.
>
> Such cosmological solutions of the gravitation-equations (with not vanishing A-constant) have been found by Mr. Gödel. It will be interesting to weigh whether these are not to be excluded on physical grounds.

However, every physical situation described by the GTR involves a distribution of matter. Without specifying the kind of matter and its placement as a distribution (which could be localized), the theory does not connect to anything physical. The mathematical world of human imagination is far vaster than the physical world, understandably. Believing that every configuration generated in the mathematical world has a counterpart in the physical world is naive. While it is true that every real physical world can be represented mathematically, the converse faith is grossly misplaced.

In our Universe, the physical conditions are more or less known and there is a manifest universal time. Every pair of events in this Universe has a uniquely determinable causal order. Then, the possibility of undetermined causality that Einstein referred to, does not happen. Whether there are other Universes where such things happen is a mute point. With our firmly founded understanding of the nature of time, we can now assert that time is not a huge serpent with a tail and (arrow) head, capable of coiling and biting its own tail. Time is an accumulated series of 'nows', all existing in the same location now, as a number of accumulated elementary durations. A 'now' cannot be compared to another that is separated in the past or future because such an entity does not exist. A 'now' here can be compared with only another 'now' elsewhere, near or far, by communication.

10.7 The Dimensions of Time

Mathematics is our medium and language for the representation of the physical world, which is constructed from the experience of the familiar physical world, assimilated over thousands of years. There is no apparent limit to the constructions of imagination in the abstract mathematical world. Some may even be useful to deal with the real physical world.

A perennial question asked by physicists and philosophers alike, and indeed even by the common man, is regarding the number of dimensions of space: is it just three as we experience, or could it be more? Surprisingly, many do not recognize that we perceive and sense only matter and its dimensions, and not space. What is interpreted as space is just extended matter. Therefore, the question about the number of dimensions of space is really the question of the number of dimensions of matter. Seen this way, much of the speculative work on this question seems baseless and wasted.

The representation of time on the quantitative number line, and ascribing a Cartesian reality to it as a coordinate in the fashion of space, spurred the question, almost by academic habit, about the number of dimensions of time. When time is blindly viewed as a coordinate, such a question makes mathematical sense, but only as that. A robust and physically consistent notion of time, closely linked to its measureability, answers in the negative about the possibility of multiple 'directions' of time. To stress the point, time itself is just the set of physical events ordered by the causality of the physical processes and interactions; these ordered events are locally felt and there is but one set of them. The same set of events does not occur in multiple copies simultaneously; definitely not at the same location, even in a speculative situation. The record of the ordered events exists as real only in the present, at a spatial location. Time does not have any existence other than as this record and as a local physical process in progress. This set that exists in matter locally is the local time. Different sets at different locations are different streams of local times which get connected only through physical interactions and cause-effect relations. This realization does not leave any room to speculate on multiple dimensions of time. The Cartesian generalization to more than one time is without any physical basis and will never connect to the description of the physical world.

One could identify different 'streams of time'. This is different from the notion of the dimensions of time. The realization that time is a local quantity, in whichever form it is experienced and measured, implies that there are infinite streams of time, one for each spatial location (as in Fig. 10.2). We have used the term 'stream' here only in the conventional sense of continuity of the present with the past and the future, of physical processes in matter. However, there is no physical sense in the notion of time where there is no matter. Since there is matter distributed almost everywhere, we will assume for simplicity that there is indeed matter everywhere, with continuity, in some form or other. If there are no physical interactions that act between spatially separated bodies, these different streams of time will never link with each other in cause-effect relations. The interaction causality is at the basis of the notion of comparison of time, simultaneity, etc. Then the relative ordering becomes relevant and physically well defined. For every observer and individual, there is only the local time, and numerous such local times get connected and are observable through transmitted signals (interactions). However, since the connection is through a signal, like the light that originates in the past physical events, the present at some location is connected only to the pasts of other streams at other locations (Fig. 10.6).

Figure 10.6 The different streams of local times, from the Earth to the far Universe, all connected to our present by light. The time in the space station (seen as a streak) is in the 'past' by 1.4 ms, and the time in Jupiter (the bright blob nearby) is 40 min in the past. The Milky Way stars in our present are thousands of years in the past of the cosmic time. Some of the stars we perceive do not even exist 'there' now! The physical past is irretrievably replaced by the present at each location (background photo credit: NASA/Bill Dunford).

It is clear that there is only one temporal dimension, if we want to treat time as a coordinate that is to be represented in a geometric generalization of the three-dimensional matter and space. All proposals with hypothetical multiple dimensions of time will remain as speculations without any counterpart in the physical world.

10.8 Can Time End?

A traversible and accessible space can be limited in its extension. *A space without boundaries and ends is imaginable only either as an infinite space or as a space closed on itself.* We do not yet know whether the Universe is finite or infinite. Both possibilities have unsolved conceptual and computational problems. A question, arising in analogy with the questions on space, is whether time has its limits or boundaries.

We discussed the mathematical possibility of the closed loop of time, which was equivalent to travelling to join the 'past' while travelling forward monotonically along time. The physical notion of time we have been adhering to does not admit this possibility. Some theoretical physicists have been wondering about, and sometimes have been alarmed of, the possibility of the end of time. This concern arises in the speculative situations called 'singularities' of space and time. These are the points of divergence of the curvature of space and time to an infinite value, as described by the GTR. In reality, they are just the locations where the gravitational tidal forces would go to infinity. This happens when the density of matter diverges to infinity. At once, we realize that one is not really talking about a real-world situation, in which matter is atomic, and there is no scope for the matter density to become truly infinite. Even if one accepts the mathematical possibility, it has no implication to time, except as a *local divergence* in matter and the termination of its physical evolution.

It is obvious that for each individual, there is an effective end of time. All local times are finite, linked to the local existence and evolution of matter. However, globally considered, matter is conserved. With our understanding of the real nature of time, we easily realize that the speculation of an end of global time is valid only in the conjunction with the end of all physical processes, or with the vanishing of all matter.

10.9 Summary

In this chapter, a general discussion of the 'arrows of time' was taken up. Though the contents were admittedly limited and incomplete, the perspective was novel, with a committed adherence to the physical view that equates the physical time to the evolving local physical processes in matter. We discussed the thermodynamic arrow of time, the cosmological arrow of time, the quantum mechanical arrow of time, and the psycho-physical arrow of time. I discussed the modern wisdom that traces the asymmetry and the apparent direction of the arrow of time to the asymmetry in the initial conditions; a relatively small number of space-time configurations of a multi-particle system corresponds to the starting point, whereas an enormously larger equivalent configurations (possibilities) are available to make a final outcome. Such arrows of time are quantitatively described in terms of the steady increase of the 'entropy' of the multi-particle physical system. The entropy signifies the incoherent, or unusable, energy that is shared between the constituents of the ensemble.

The cosmological arrow of time is related to the steady expansion of the Universe. This is then directly related to the gravitational interaction of the matter in the Universe. The special role of gravity and the irreversible evolution of an aggregate of matter into a black hole were discussed in the context of the question of the reversibility of time. Here, the irreversibility of the evolution of the matter configuration is absolute, even though the interaction is considered mathematically time symmetric. The phenomenon of black hole formation makes gravity very different from other known interactions, in the context of the nature of time.

The quantum mechanical arrow of time is also an effective direction in the physical processes, but its fundamental irreversibility, even in the case of the dynamics of a single particle, arises in the process of measurement, or observation. The nature of the quantum measurement is still an open problem.

The psychological arrow of time is our most familiar directed 'flow of time'. Its relation to the physical notion of memory and its spatial and local existence in matter were mentioned. I did not discuss much about the non-linear psychological time and its perceived direction because of the complexity of the subject and the obvious difficulties in simply relating it to the linear physical time.

I argued that the arrow of time is not related to any inherent direction in the elemental physical time that is equated to the fundamental physical processes. However, a formal mathematical relation, in which the abstract time is treated as a 'coordinate' and the general notion of energy as the 'generator' of translations along the time-coordinate, was mentioned ('bow for the arrow'). Then I discussed the true nature of time, in which I argued that only the 'present' (NOW) is in physical existence, that too locally. Past is a mere record of simultaneities or coincidences between a conventional clock and the physical processes, a progressively increasing numeral record, all existing locally in space, and only in the present. A sense of global time is established by communication and physical interactions between spatially separated locations.

This robust and consistent physical picture helped us in answering many deep questions about time. The open issues that we addressed and answered include time reversal, time travel, antiparticles as time-reversed particles, closed loops of time, and the dimensions of time. I discussed the crucial qualitative difference in the description of the antiparticles when their mass is explicitly included, along with their charges. All speculations about time travel are put to rest once we realize the physical nature of time, as existing only as current physical processes, now and here.

Chapter 11

The Last Ticks

We have come a long way together, exploring the links between gravity and time. The detailed discussion established the determining role of cosmic gravity in the physical modifications of time, and indeed in all physical processes. A satisfying consistency was achieved when we realized that time itself is in fact some evolving physical processes. We have explored only those physical links between cosmic gravity and physical processes that are relevant for our topic of time and clocks. The theory of Cosmic Relativity is spread over several canvases, touching every sphere of physics, because all of our physics is done in this unique Universe. The references listed discuss these various topics.

11.1 Time and Quantum Gravity

There is one important topic concerning time and gravity that is noticeably absent in the book – a discussion on the elusive theory of Quantum Gravity! That is not an omission. I was focused on the provable and demonstrable relations between gravity and time. All the topics discussed in this book so far are firmly supported by experimental evidence and empirically verifiable arguments. The

Gravity's Time
C. S. Unnikrishnan

ISBN 978-981-4877-72-5 (Hardcover), 978-1-003-25656-4 (eBook)
www.jennystanford.com

clear grasp of the conceptual and practical aspects of physical time and its modifications due to gravity and motion have been the recurring themes. Speculations and future theories had no place in that context.

However, certain brief remarks about time and the theory of quantum gravity might be relevant. Quantum gravity is the theoretical hope of describing the gravitational interaction consistently including the features of quantum mechanics (Smolin, 2001). This goal has been pursued very seriously for several decades without any concrete outcome. The belief that such a theory could be constructed and indeed necessary comes from the examples of quantum electrodynamics and the 'quantized' theories of other interactions. Those theories were needed to deal with the physical processes in the atomic and the subatomic world. They were prompted and supported by many empirical facts and experimental results. In contrast, empirical clues for quantum gravity have been totally absent and nobody has been able to advocate a real need for a theory of quantum gravity. One obvious reason is that the gravitational effects are too weak to be manifest in the atomic world. When we come to the macroscopic world where gravity starts to become prominent, the characteristic stochasticity of quantum mechanics becomes negligible and irrelevant. The only topic for which one could argue for an essential need for quantum gravity is perhaps the very early Universe and cosmology near the origin of the Universe. However, there is a fundamental problem in the sense that quantum mechanics deals with probabilities of an ensemble of systems, whereas there is only one Universe as the physical system. How to make quantum sense out of a single copy of the physical system is not clear. One can see that contrary to the general impression, quantum gravity may very well be a theory without any observable consequence, except formal significance. In any case, there is no theory of 'Quantum General Relativity' yet.

Time is a critical conceptual factor in quantum gravity. I mention just two aspects. In 1900, M. Planck introduced the fundamental units for physical quantities based on the fundamental constants G, c and $\hbar$, now called the Planck units. For the basic quantities of length, time and mass these units are the Planck length $l_P \equiv \sqrt{\hbar G/c^3} \simeq 1.62 \times 10^{-35}$ m, the Planck time $t_P \equiv l_P/c = \sqrt{\hbar G/c^5} \simeq$

5.39×10^{-44} s, and the Planck mass $m_P \equiv \sqrt{\hbar c/G} \simeq 2.18 \times 10^{-8}$ kg. They are not very useful in practical measurements, but have been useful in limited theoretical contexts. There are a lot of speculations knitted around these units because the fundamental constants G, c and $\hbar$ refer to gravity, relativity and quantum mechanics. For some researchers, the Planck units define the numerical scales of the quantized relativistic gravitational phenomena, which would be described by the theory of quantum gravity. There is one obvious catch though; while the units for length and time involve unimaginably tiny measures where a speculation can live without scrutiny, the Planck mass is not really microscopic. In fact, it is macroscopic and large compared to a mass in the atomic scale. But nothing spectacular happens in either gravity or quantum mechanics when we deal with such masses (like a tiny glass bead). In any case, many physicists believe that these are scales of quantum gravity and that quantum gravitational effects should manifest at those spatial and temporal scales. However, there are no concrete ideas on what might be those effects. People who naively ascribe physicality to empty space and time without matter even think that space and time themselves are discrete and fragmented in some way, on such scales. Such speculations can be put on a mathematical basis, as always possible, but their physical support is nonexistent.

The second aspect concerns the role of time as the reference for the physical evolution (dynamics). Since the theory of gravity is relativistic, where space and time are treated as equivalent, there are fundamental unresolved issues of extracting a unique physical time to measure the evolution of physical states. Another deep issue is about the physical entities whose time evolution is sought in quantum gravity. The dynamical variables are the metric of the space-time, or in effect space and time themselves! Then, there is no 'time' left to mark the physical evolution of the space-time. From my point of view, these problems are diversions without any physical basis. Such problems arise because the theory is formulated in empty space with its matter-less time, because of its foundational basis built from the STR and the GTR. Further, any approach to a theory of quantum gravity presupposes the physical existence of a 'space-time'. We have already seen from many points of view that the concept of a four-dimensional space-time is a metaphysical unobservable entity, without a physical meaning except in relation to the matter and its evolution. Further, we have discussed how the theoretical assumption that the 'axis of time' extends indefinitely from the present to both the past and the future is physically unjustified. Once we acknowledge the physical presence of the matter-filled Universe, the universal time available from the evolving Universe, and the gravitational effects of the cosmic matter on time, the entire perspective changes.

There are very strong indications that the quest through generations for the theory of quantum gravity is mostly on the wrong tracks (Unnikrishnan, 2010). The only way a formal theory can be completed is by acknowledging the gravity of the cosmic matter and its universal time, and by recognizing the unphysical nature of treating the empty space and its time as the dynamical variables. In any case, we need not discuss it further because it is not relevant for the physics of time and its modifications.

11.2 The Future of Time

The future of time will be based on our understanding of time as Gravity's Time. This is inevitable if we have to honestly deal with time. The existence of matter in the Universe and its gravity cannot be denied or ignored. Then, one cannot continue to be blind to its plain effect on time, and on real clocks. As the technological need for more and more precise time increases, the factual situation that the Universe is the master frame that defines and controls the universal time and simultaneity will become more and more clear and evident. The atomic time and its gravitational as well as motional modifications are becoming accessible and important for most people, through their navigational needs and their continuously monitored existence. Then, the true nature of time and the correct theoretical basis for its description will eventually and inevitably become common knowledge. It will be a sociologically bizarre situation if the common people are fully aware of the physics of time and its gravitational basis, while the community of physicists are trapped in a time warp of beliefs that originated in a-century-old scriptures!

Looking back, one will realize that just by acquiring two basic factual insights – that time is nothing but the evolving states of matter, inseparable and localized in matter, and that the gravity of the cosmic matter is the determining factor for the modifications of time and the propagation of light – we became capable and competent to understand and deal with even the most intricate issues concerning the physical time. This is indeed a very satisfying position.

Every conceptual and practical question about time, duration, and simultaneity finds a reliable and accurate answer when time is understood as in Cosmic Relativity, as Gravity's Time.

Bibliography

Abbott, B. P. (2017). Gravitational waves and gamma-rays from a binary neutron star merger: GW170817 and GRB170817a, *Astrophys. J. Lett.* **848**, p. L13.

Aristotle (1991). *Physics* (Princeton University Press).

Ashby, N. (2002). Relativity and the global positioning system, *Phys. Today* **55**, pp. 41–47.

Ashby, N. (2003). Relativity in the global positioning system, *Living Rev. Relativity,* **6**, 1, pp. 1–42.

Bell, J. S. (1964). On the Einstein–Podolsky–Rosen paradox, *Physics* **1**, pp. 194–200.

Bergson, H. (1922). *Durée et Simultanéité: à propos de la théorie d'Einstein* (Duration and Simultaneity: With reference to Einstein's theory, English translation by L. Jacobson, (Bobbs-Merril Company Inc. USA, 1965)).

Brush, S. G. (1992). How cosmology became a science, *Sci. Am.*, pp. 62–70.

Champeney, D. C., Isaak, G. R., and Khan, A. M. (1963). An aether drift experiment based on the Mössbauer effect, *Phys. Lett.* **7**, pp. 241–243.

Chou, C. W., Hume, D. B., Rosenband, T., and Wineland, D. J. (2010). Optical clocks and relativity, *Science* **329**, pp. 1630–1633.

Ciufolini, I., and Wheeler, J. A. (1995). *Gravitation and Inertia* (Princeton University Press).

Einstein, A. (1905). On the electrodynamics of moving bodies, *Ann. Phys.* **17**, pp. 891–921.

Einstein, A. (1911). On the influence of gravitation on the propagation of light, *Ann. Phys.* **35**, pp. 898–908.

Einstein, A. (1914). Physical foundations of a theory of gravitation, *Naturforschende Gesellschaft in Zürich. Vierteljahrsschrift* **58**, pp. 284–290.

Einstein, A. (1918). Dialogue about objections to the theory of relativity, *Die Naturwissenschaften* **6**, p. 697, the English translation of the original German version is available in the translated companion volume of The Collected Papers of Albert Einstein, Volume 7, The Berlin Years: Writings, 1918-1921, pp. 66-75 (Princeton University Press, Princeton, NJ, USA).

Einstein, A. (1920). *Relativity: The Special and General Theory* (Methuen & Co. Ltd, London), translated by R. W. Lawson.

Einstein, A. (1936). Physics and reality, *J. Franklin Inst.* **221**, pp. 349–382.

Einstein, A. (1948). Quantum mechanics and reality (in German), *Dialectica* **2**, pp. 320–324.

Einstein, A. (1949). *Albert Einstein: Philosopher-Scientist,* Schilpp, P. A. (Ed.), chap: Remarks to the Essays Appearing in this Collective Volume (MJF Books, NY, USA).

Einstein, A. (1956). *The Meaning of Relativity* (Methuen, London).

Einstein, A., Podolsky, B., and Rosen, N. (1935). Can quantum mechanical description of physical reality be considered complete? *Phys. Rev.* **47**, pp. 777–780.

Feynman, R. P. (1949). The theory of positrons, *Phys. Rev.* **76**, pp. 749–759.

Galileo, G. (1638). *Dialogues Concerning Two New Sciences* (William Andrew Publishing, NY, USA).

Gruber, T., Eicker, A., and Flechtner, F. (eds.) (2019). Special issue on *Remote Sensing by Satellite Gravimetry*, MDPI Remote Sensing.

Hafele, J. C. (1971). Performance and results of portable clocks in aircraft, Report from Washington University, St. Louis at the Department of Defense (DOD) Precise Time and Time Interval (PTTI) Strategic Planning Meeting, 16–18 November 1971, Washington, DC.

Hafele, J. C., and Keating, R. E. (1972). Around-the-world atomic clocks: Observed relativistic time gains, *Science* **177**, pp. 168–170.

Hamilton, W. R. (1833). On a general method of expressing the path of light, and of the planets, by the coefficients of a characteristic function, *Dublin Univ. Rev. Q. Mag.* **1**, pp. 795–826.

Hamilton, W. R. (1835). Second essay on a general method indynamics, *Philos. Trans. R. Soc.* **part 1**, pp. 95–144.

Heidegger, M. (1927). *Being and Time* (Blackwell Publishers Ltd., UK).

Hubble, E. (1929). A relation between distance and radial velocity among extra-galactic nebulae, *PNAS* **15**, pp. 168–173.

Jones, T. (2000). *Splitting the Second: The Story of Atomic Time* (Institute of Physics Publishing).

Kennedy, R. J., and Thorndike, E. M. (1932). Experimental establishment of the relativity of time, *Phys. Rev.* **42**, pp. 400–418.

Krisher, T. P. (1980). Test of the isotropy of the one-way speed of light using hydrogen-maser frequency standards, *Phys. Rev.* **42**, pp. 731–734.

Landes, D. S. (2000). *Revolution in Time: Clocks and the Making of the Modern World* (Harward University Press).

Langevin, P. (1911). L'évolution de l'espace et du temps, *Scientia* **5**, p. 31.

Lopez, O, et al. (2015). Frequency and time transfer for metrology and beyond using telecommunication network fibres, *C. R. Phys.*, **16**, pp. 531–539.

Lorentz, H. A. (1904). Electromagnetic phenomena in a system moving with any velocity smaller than that of light, *Proc. R. Neth. Acad. Arts Sci.* **6**, pp. 809–831.

Mach, E. (1919). *The Science of Mechanics* (The Open Court Publishing Co., Chicago).

Maxwell, J. C. (1954). *A Treatise on Electricity and Magnetism* (Dover Publications, Inc. NY, USA).

McCarthy, D. D., and Seidelmann, K. P. (2009). *TIME: From Earth Rotation to Atomic Physics* (WILEY-VCH Verlag GmbH & Co.).

Michelson, A. A. (1904). Relative motion of earth and aether, *Philos. Mag.* **8**, pp. 716–719.

Michelson, A. A. (1927). *Studies in Optics* (Dover Publications).

Michelson, A. A., and Gale (1925). The effect of the earth's rotation on the velocityof light, ii, *Astrophys. J.* **61**, pp. 140–145.

Michelson, A. A., and Morley, E. W. (1887). On the relative motion of the earth and the luminiferous ether, *Am. J. Sci.* **34**, pp. 333–345.

Minkowski, H. (2012). *Space and Time: Minkowskis Papers on Relativity* (Minkowski Institute Press, Montréal).

Morris, M. S., and Thorne, K. S. (1988). Wormholes in spacetime and their use for interstellar travel: A tool for teaching general relativity, *Am. J. Phys.* **56**, pp. 395–412.

Newton, I. (1846). *Principia Mathematica (The Mathematical Principles of Natural Philosophy)* (Daniel Adee, NY, USA).

Nordmann, C. (1922). Einstein expose et discute sa théorie, *Revue des Deux Mondes* **9**, pp. 129–166, English translation by members of the La Rouche movement in 21st Century Science and Technology, Summer 2011.

Planck, M. (1915). *Eight Lectures on Theoretical Physics* (Columbia University Press, NY, USA).

Poincaré, H. (1898). La mesure du temps, *Revue de métaphysique et de morale* **6**, pp. 1–13.

Poincaré, H. (1900). La théorie de Lorentz et le principe de réaction, *Archives Néerlandaises des Sciences Exactes et Naturelles, 2ème série* **5**, pp. 252–278.

Poincaré, H. (1904). L'état actuel et l'avenir de la physique mathématique, *Bulletin des sciences mathématiques* **28**, pp. 302–324.

Poincaré, H. (1905). Sur la dynamique de l'électron, *Comptes Rendus de l' Acad. des Sc.* **14**, pp. 1504–1508.

Pound, R. V., and Rebka, G. A. (1960). Apparent weight of photons, *Phys. Rev. Lett.* **4**, pp. 337–341.

Price, R. H. (1982). General relativity primer, *Am. J. Phys.* **50**, pp. 300–329.

Reichenbach, H. (1971). *The Direction of Time* (The University of California Press, Berkeley).

Ruderfer, M. (1960). First order terrestrial ether drift using the Mössbauer radiation, *Phys. Rev. Lett.* **5**, p. 191.

Ruderfer, M. (1961). Erratum: First order terrestrial ether drift using the Mössbauer radiation, *Phys. Rev. Lett.* **7**, 361.

Ryden, B. (2002). *Introduction to Cosmology* (Cambridge University Press).

Sagnac, G. (1913a). L'éther lumineux démontré par l'effet du vent relatif d'éther dans un interféromètre en rotation uniforme, *C. R. Acad. Sci. Paris* **157**, pp. 708–710.

Sagnac, G. (1913b). Sur la preuve de la réalité de l'éther lumineux par l'expérience de l'interférographe tournant, *C. R. Acad. Sci. Paris* **157**, pp. 1410–1413.

Sciama, D. W. (1953). On the origin of inertia, *MNRAS* **113**, pp. 34–42.

Smolin, L. (2001). *Three Roads to Quantum Gravity* (Basic Books).

Taylor, E. F., and Wheeler, J. A. (1992). *Spacetime Physics* (J. H. Freeman and Co.).

Thompson, W., and Tait, P. G. (1879). *Elements of Natural Philosophy* (P. F. Collier and Son, NY, USA).

Unnikrishnan, C. S. (2004). Cosmic Relativity: The theory, its implications and experimental tests, ArXiv:gr-qc/0406023.

Unnikrishnan, C. S. (2005a). Correlation functions, Bell's inequalities and the fundamental conservation laws, *Europhys. Lett.* **69**, pp. 489–495.

Unnikrishnan, C. S. (2005b). On Einstein's resolution of the twin-clock paradox, *Curr. Sci.* **89**, p. 2009.

Unnikrishnan, C. S. (2009). Physics in the 'once-given' universe, in S. Kar and G. Ghosh (eds.), *Recent Advances in Theoretical Physics* (World Scientific, Singapore), pp. 99–120.

Unnikrishnan, C. S. (2010). *History of Science and Philosophy of Science Volume XIII*, chap. Quantum Gravity and the Structure of Physical Reality (Centre for Studies in Civilizations & Pearson-Longman, Delhi), pp. 409–444.

Unnikrishnan, C. S. (2011). Precision measurement of the one-way speed of light and implications to the theory of relativity, in M. C. Duffy, V. O. Gladyshev, A. Morozov and P. Rowlands (eds.), *Physical Interpretations of Relativity Theory: Proceedings* (Bauman Moscow State Technical University), pp. 505–511.

Unnikrishnan, C. S. (2014). True dynamical tests of the equivalence principle, *Int. Jl. Mod. Phys (Conf. series)* **30**, p. 1460267.

Unnikrishnan, C. S. (2018a). Experimental evidence for the gravitational basis for relativity and dynamics, in S. D. M. G. C. S. M. Wagh (ed.), *The Physical Universe* (Central India Research Institute, Nagpur, India), pp. 183–202.

Unnikrishnan, C. S. (2018b). Reconstructing quantum mechanics without foundational problems, ArXiv:1812.06088 [quant-ph].

Unnikrishnan, C. S. (2020a). A new gravitational paradigm for relativity and dynamics, and its philosophical scope, *J. Phys.: Conf. Ser.* **1466**, p. 012007.

Unnikrishnan, C. S. (2020b). The theories of relativity and Bergson's philosophy of duration and simultaneity during and after Einstein's 1922 visit to Paris, ArXiv:2001.10043 [physics.hist-ph].

Unnikrishnan, C. S. (2022). New Relativity in the Gravitational Universe (Springer, forthcoming).

Unnikrishnan, C. S. (2021b). On the unconditional validity of J. von Neumann's proof of the impossibility of hidden variables in quantum mechanics, ArXiv:2105.13996.

Unnikrishnan, C. S., and Gillies, G. T. (2002). *Pushing Gravity: New Perspectives on Le Sage's Theory of Gravitation*, chap. Constraints on Gravitational Shielding (Apeiron), pp. 259–266.

Weinberg, S. (1977). *The First Three Minutes* (Basic Books, New York).

Wells, J. W. (1963). Coral growth and geochronometry, *Nature* **197**, pp. 948–950.

Whittaker, E. T. (1954). *A History of the Theories of Aether & Electricity: The Modern Theories* (Thomas Nelson, Edinburgh-London).

Zeh, D. H. (1989). *The Physical Basis of The Direction of Time* (Springer, Berlin, Heidelberg).

Index